Prepared in cooperation with Routt County, the Colorado Water Conservation Board, and the City of Steamboat Springs

Water-Quality Assessment and Macroinvertebrate Data for the Upper Yampa River Watershed, Colorado, 1975 through 2009

Scientific Investigations Report 2012–5214

U.S. Department of the Interior
U.S. Geological Survey

Water-Quality Assessment and Macroinvertebrate Data for the Upper Yampa River Watershed, Colorado, 1975 through 2009

By Nancy J. Bauch, Jennifer L. Moore, Keelin R. Schaffrath, and Jean A. Dupree

Prepared in cooperation with Routt County, the Colorado Water Conservation Board, and the City of Steamboat Springs

Scientific Investigations Report 2012–5214

U.S. Department of the Interior
U.S. Geological Survey

U.S. Department of the Interior
KEN SALAZAR, Secretary

U.S. Geological Survey
Marcia K. McNutt, Director

U.S. Geological Survey, Reston, Virginia: 2012

For more information on the USGS—the Federal source for science about the Earth, its natural and living resources, natural hazards, and the environment, visit http://www.usgs.gov or call 1–888–ASK–USGS.

For an overview of USGS information products, including maps, imagery, and publications, visit http://www.usgs.gov/pubprod

To order this and other USGS information products, visit http://store.usgs.gov

Suggested citation:
Bauch, N.J., Moore, J.L., Schaffrath, K.R., and Dupree, J.A., 2012, Water-quality assessment and macroinvertebrate data for the Upper Yampa River watershed, Colorado, 1975 through 2009: U.S. Geological Survey Scientific Investigations Report 2012–5214, 129 p.

Contents

Figures

Tables

Conversion Factors, Abbreviations, and Datum

Multiply	By	To obtain
acre-foot (acre-ft)	1,233	cubic meter (m^3)
cubic foot per second (ft^3/s)	0.02832	cubic meter per second (m^3/s)
foot (ft)	0.3048	meter (m)
inch per year (in/yr)	25.4	millimeter per year (mm/yr)
square meter (m^2)	10.76	square foot (ft^2)
square mile (mi^2)	2.590	square kilometer (km^2)

Temperature in degrees Celsius (°C) may be converted to degrees Fahrenheit (°F) as follows:
$$°F=(1.8×°C)+32$$

Vertical coordinate information is referenced to the North American Vertical Datum of 1988 (NAVD 88).

Horizontal coordinate information is referenced to the North American Datum of 1983 (NAD 83) or the North American Datum of 1927 (NAD 27).

Altitude, as used in this report, refers to distance above the vertical datum.

Water year, as used in this report, refers to the period October 1–September 30 and is designated by the year in which it ends.

Specific conductance is given in microsiemens per centimeter at 25 degrees Celsius (µS/cm).

Concentrations of chemical constituents in water are given in either milligrams per liter (mg/L) or micrograms per liter (µg/L). A value of 1 mg/L equals 1 part per million; a value of 1 µg/L equals 1 part per billion.

Additional Abbreviations, Acronyms, and Symbols

AMLE	adjusted maximum likelihood estimation
ANC	acid neutralizing capacity
$CaCO_3$	calcium carbonate
CDOA	Colorado Department of Agriculture
CDPHE	Colorado Department of Public Health and Environment
col/100 mL	colonies per 100 milliliters
DM	daily maximum temperature
E. coli	*Escherichia coli*
HCO_3^-	bicarbonate
HH	human health
MCL	maximum contaminant level
MWAT	maximum weekly average temperature
N	nitrogen
NH_3	un-ionized ammonia
NH_4^+	ammonium
P	phosphorus
ROS	regression of ordered statistics
SMCL	secondary maximum contaminant level
STORET	STOrage and RETrieval
TVS	table value standard
USEPA	U.S. Environmental Protection Agency
USGS	U.S. Geological Survey
UYRW	Upper Yampa River watershed
WS	water supply
<	less than

Page intentionally left blank

Water-Quality Assessment and Macroinvertebrate Data for the Upper Yampa River Watershed, Colorado, 1975 through 2009

By Nancy J., Bauch, Jennifer L. Moore, Keelin R. Schaffrath, and Jean A. Dupree

Abstract

A study was initiated in 2009 by the U.S. Geological Survey (USGS), in cooperation with Routt County, the Colorado Water Conservation Board, and the City of Steamboat Springs, to compile and analyze historic water-quality data and assess water-quality conditions in the Upper Yampa River watershed (UYRW) in northwestern Colorado. Water-quality data for samples collected by federal, state, and local agencies for various periods from 1975 through 2009 were compiled and assessed for streams, lakes, reservoirs, and groundwater in the UYRW, including the Elkhead Creek subwatershed and the Yampa River watershed that is upstream from Elkhead Creek. For selected physical-property and chemical-constituent data for samples collected from surface-water sites and groundwater wells in the UYRW, this report: (1) characterizes available data through statistical summaries, (2) analyzes the spatial and temporal distribution of water-quality conditions, (3) identifies temporal trends in water quality, where possible, (4) provides comparisons to federal and state water-quality standards and recommendations, and (5) identifies factors affecting the quality of water. In addition, the availability and characteristics of macroinvertebrate data collected in the UYRW are described.

Water-quality data were compiled for 211 stream sites located throughout much of the watershed. A total of 5,861 stream-water samples with data for physical properties, dissolved solids, major ions, nutrients, trace elements, uranium, coliform bacteria, and suspended-sediment concentrations collected from 1975 through 2009 were analyzed. The amount of data collected varied by year and site.

Values of specific conductance and values and concentrations for other water-quality constituents depended, in part, on the geology underlying a stream's drainage basin. Specific conductance values were lower in areas with igneous and metamorphic rocks than in areas with sedimentary rocks. Water temperature for main-stem Yampa River sites increased in a downstream direction; water temperatures in one Yampa River site exceeded Colorado Department of Public Health and Environment (CDPHE) water-quality standards for protection of aquatic life. Streams with drainage basins underlain by igneous and metamorphic rocks tended to have softer water and a lower capacity to neutralize inputs of acidic water than streams with drainage basins underlain by sedimentary rocks. The spatial distribution of dissolved solids and major ions was similar to that for specific conductance. Values and concentrations also tended to be lower during snowmelt runoff than other times of the year.

Many concentrations of dissolved nitrite and nitrate and unfiltered total ammonia in stream-water samples were less than laboratory detection levels; however, about 14 percent of the samples with unfiltered total phosphorus data had concentrations that exceeded federal recommendations. A statistically significant upward trend in unfiltered total phosphorus concentrations at a Yampa River site in Steamboat Springs may reflect population growth and land-use changes that have occurred upstream from the site.

About two-thirds of the concentration data for many trace elements in stream-water samples also were less than laboratory detection levels. Maximum concentrations of the various trace elements occurred in main-stem Yampa River subwatersheds and seemed to depend on the lithology of the rocks underlying a subwatershed. Some sites were not in attainment of state aquatic-life standards for dissolved copper, dissolved selenium, and total recoverable iron and water-supply standards for dissolved iron and manganese.

Concentrations of the bacterium *Escherichia coli* in five stream samples collected from 1994 through 2003 were greater than the state recreation standard. High values could be due to recreational users of a stream, wildlife, and (or) livestock.

Water-quality data for Lake Elbert, Long Lake Reservoir, Stagecoach Reservoir, Steamboat Lake, and Elkhead Reservoir for various periods of time since 1985 were summarized or analyzed. Lake Elbert and Long Lake Reservoir were very dilute and have little capacity to neutralize inputs of acid. Anoxic conditions (dissolved oxygen concentrations less than 0.5 milligrams per liter) at depth were indicated for Stagecoach Reservoir and Steamboat Lake during July 2006 and for Elkhead Reservoir on some days from July 1995 through August 2001. The trophic status of Elkhead Reservoir ranged from oligotrophic to eutrophic.

A total of 816 groundwater-quality samples collected from 328 wells during 1975 through 1989 and 1998 were analyzed for this study. The sampled wells are concentrated in the middle latitudes of the UYRW. About 66 percent of the wells with water-quality data were sampled only once. Samples were collected from wells that tap aquifers in 12 named geologic units and geologic units that have not been identified. More groundwater samples were collected from wells completed in the unknown geologic units, Mesaverde Group, and terrace alluvium than from wells completed in other geologic units.

Analysis of groundwater data for physical properties indicates that specific conductance was lower in samples collected from igneous and metamorphic rocks and sedimentary rocks of nonmarine origin than sedimentary rocks of marine or marine-nonmarine origin. Values of pH not meeting CDPHE groundwater standards were most commonly collected from the flood-plain alluvium and unknown geologic units. The CDPHE water-supply standard for dissolved sulfate in groundwater was exceeded in concentrations from about one-half of the samples collected, most commonly for samples collected from the terrace alluvium, Mesaverde Group, and unknown geologic units.

All dissolved nitrite concentrations were well below CDPHE maximum contaminant level (MCL) for nitrite in groundwater, and fewer than 5 percent of dissolved nitrate plus nitrite concentrations were greater than the CDPHE MCL for nitrate. Almost all dissolved and unfiltered total phosphorus concentrations in groundwater samples were less than 0.1 milligram per liter.

More than 80 percent of samples collected for some trace elements had concentrations that were less than or equal to laboratory detections levels. Exceedances of CDPHE water-quality standards were rare for most trace elements. Only one or two samples each had concentrations of dissolved arsenic, beryllium, copper, lead, molybdenum, and selenium that exceeded human-health standards, MCLs, and (or) agricultural-use standards. Less than 4 percent of the samples had dissolved cadmium concentrations that exceeded the CDPHE MCL for cadmium in groundwater. Agricultural-use standards were exceeded for about 13 percent of dissolved boron samples and less than 1 percent of dissolved zinc samples. Exceedances of CDPHE secondary maximum contaminant levels were detected for about 10 percent of dissolved iron samples and more than one-half of dissolved manganese samples. These samples with exceedances were collected from the Mancos Shale, valley-fill deposits, and unknown geologic units (dissolved iron) and terrace alluvium, Mesaverde Group, and unknown geologic units (dissolved manganese).

Macroinvertebrate community and population data were available for 66 stream sites in the UYRW for various periods from 1975 through 2008. A summary of results from one study of Yampa River sites in Steamboat Springs indicates that changes observed in community characteristics between 2005 and 2008 may be due to upstream effects or large-scale environmental factors rather changes in water quality within the stream reach.

Synthesis of water-quality data indicates that the values and concentrations of many physical properties and constituents in surface-water samples for the UYRW are likely controlled primarily by geology, streamflow, and land use. The quality of groundwater in the UYRW is a function of various physical and geochemical processes, including precipitation, the depositional environment of the aquifer sediments, type of sediments that groundwater moves through, dissolution of soluble minerals in rocks and soils, and ion exchange reactions. Constituents that are issues of concern for aquatic life, human health, or suitability of water for various uses include those on the CDPHE 2012 303(d) (federal Clean Water Act Section 303d) list of impaired waters or monitoring or evaluation list for surface water. Other constituents in stream water or groundwater that are or could be issues of concern include those that commonly have concentrations that exceed standards or that could affect technical qualities of water. This could include unfiltered sulfate, unfiltered total phosphorus, and dissolved copper in stream water; pH, unfiltered nitrate plus nitrite, and dissolved copper in groundwater; and hardness, dissolved solids, iron, and manganese in stream water and groundwater. Analysis of stream-water and groundwater data for changes in water quality over time was limited because of the absence of long-term data collection in the Upper Yampa River Watershed.

Introduction

The Yampa River, the largest primarily free-flowing tributary to the Colorado River in the Upper Colorado River basin (Blinn and Poff, 2005), is a highly valued resource known for its biological diversity, largely unaltered natural condition, and generally high water quality. The Upper Yampa River watershed (UYRW) (fig. 1), defined for this study as the Elkhead Creek subwatershed and the Yampa River watershed upstream from Elkhead Creek in northwestern Colorado, is undergoing increased land and water development to support growing municipal demands, recreational tourism, and second-home development that present water-quality challenges. A plan for the Yampa River watershed was developed in 2002 for the Colorado Department of Public Health and Environment (CDPHE) and the Yampa River Basin Partnership (Montgomery Watson Harza, 2002). The main goals of the plan were to address water-quality concerns and provide for the maintenance of high quality water in the Yampa watershed. State and local stakeholders that rely on and manage the water resources of the watershed are interested in an assessment of water quality to aid in the preservation and management of the UYRW. Stakeholders have specifically expressed a need for a compilation and evaluation of the available historic UYRW water-quality data to assess the effects of growth and associated land-use changes on water quality, identify spatial and temporal gaps within available water-quality data, and evaluate spatial and temporal trends in water quality. In 2009, the U.S. Geological Survey (USGS), in cooperation with Routt County, the Colorado Water Conservation Board, and the City of Steamboat Springs, initiated a study to compile

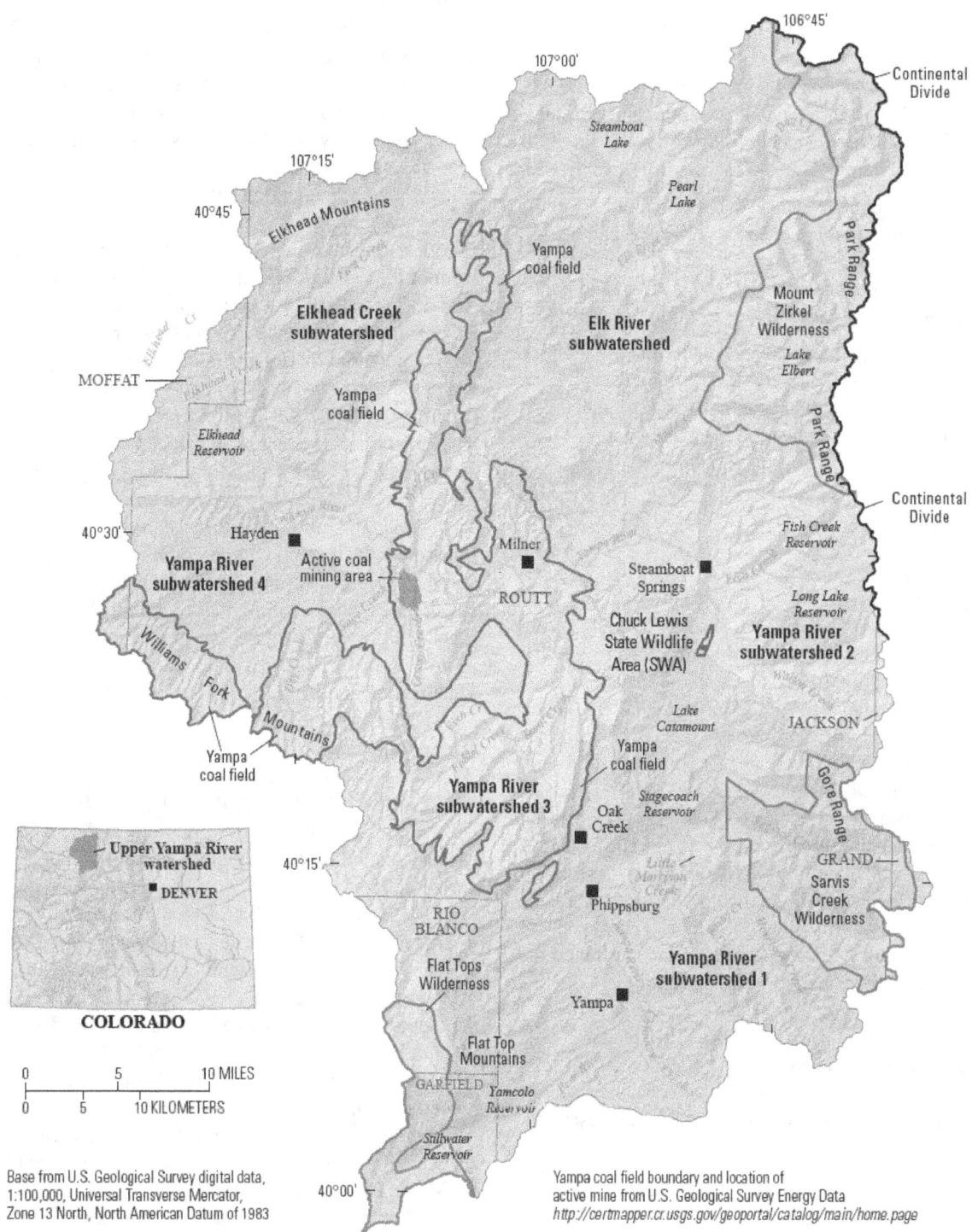

Figure 1. Location of the Upper Yampa River watershed, Colorado.

water-quality and macroinvertebrate data and assess water-quality conditions in the UYRW. Specific objectives of the study were to:

- Develop and maintain a web-accessible water-quality database that provides agencies, researchers, consultants, and interested stakeholders with equal access to historic and current water-resources information;

- Evaluate available water-resources data for uniformity and suitability to meet the needs of water- and land-resource managers and decision makers as well as the public and other stakeholders, and perform and publish an assessment of water-resource conditions;

- Design and implement regional water-quality monitoring strategies to effectively fill identified data gaps by reducing duplication of effort while still meeting a broad base of data-collection objectives; and

- Upon implementation of the monitoring program, periodically assess the new data to update factors affecting water-resource conditions.

Purpose and Scope

The purpose of this report is to describe and provide an analysis of water-quality data collected in the UYRW by federal, state and local agencies from 1975 through 2009. Water-quality data include physical properties (specific conductance, pH, water temperature, dissolved oxygen, hardness, and acid neutralizing capacity), streamflow, dissolved solids, major ions, nutrients, trace elements, uranium, coliform bacteria, and (or) suspended sediment for streams, lakes, reservoirs, and (or) groundwater wells. Methods used to analyze these data are described. For selected physical properties and water-quality constituents in the UYRW, this report: (1) characterizes available data through statistical summaries, (2) evaluates the spatial and temporal distribution of water quality, (3) identifies temporal trends in water quality, where possible, (4) provides comparisons to federal and state water-quality standards and recommendations, and (5) identifies factors affecting the quality of water. In addition, the availability and characteristics of macroinvertebrate data collected in the UYRW are described.

Study Area

The UYRW drains approximately 1,800 square miles of the Yampa River watershed west of the Continental Divide in northwestern Colorado (fig. 1) (U.S. Geological Survey, 2010). The boundaries of the watershed extend from the Williams Fork and Flat Top Mountains in the southwestern and southern portions of the watershed, respectively, to the Gore and Park Ranges and the Continental Divide to the east and to the Elk River and Elkhead Creek drainages to the north and west, respectively. Altitudes in the watershed range from more than 12,000 feet (ft) (above North American Vertical

Datum of 1988) in the Flat Top Mountains and Park Range to 6,400 ft near the confluence of the Yampa River with Elkhead Creek west of the town of Hayden. The UYRW is almost entirely contained within Routt County, with small portions in Grand, Garfield, Jackson, Moffat, and Rio Blanco Counties.

Human activity in the UYRW began at least 1,000 years ago when the Native Americans used the Yampa River valley for summer hunting (*http://yampavalley.info/centers/history_%2526_genealogy*, accessed June 2012). Trappers of European origin came to the valley around 1820. Development of the valley was sparked with the 1861 discovery of gold at Hahns Peak (not shown on fig. 1) in the northern Elk River subwatershed. The vast coal resources in the region and the potential for grazing lands were noted by Ferdinand Hayden during a 1874 survey across northwestern Colorado (Mehls and Mehls, 1991). Initially, coal was mined for local use; larger-scale production began in 1909 with improved transportation to and from Steamboat Springs. For the greater part of the past century, ranching, including hay and wheat production, and mining were the economic bases of the valley. More recently, recreational-based tourism (skiing, fishing, hunting, rafting, camping, among others) and second-home development became important economic drivers. These accounted for approximately 45 percent of the total jobs in Routt County during 2008 (State Demography Office, 2010a). The Steamboat Springs Ski Resort attracts skiers from around the world. During the 2008–2009 ski season, the number of skier days (skier day is one individual visiting a ski area for skiing or snowboarding for any part of one day) at Steamboat Ski Resort was 959,603 (C. Bannister, Colorado Ski Country USA, oral commun., 2010). Agriculture and coal mining each accounted for only about 2.5 percent of the total jobs during 2008.

The population of Routt County during 2008 was estimated to be 22,931 (State Demography Office, 2010b). The largest township is Steamboat Springs (12,280); and the next largest is Hayden (1,672). About 34 percent of the population in the county lived in unincorporated areas. From 2000 through 2008, the population of Routt County grew by more than 16 percent and was largely driven by recreation-related tourism.

The dominant land cover in the UYRW is forest, which accounts for about 57 percent of the total land area (fig. 2) (LaMotte, 2008). Other prominent land covers are shrub/scrub (about 26 percent) and grassland/pasture (about 13 percent). Barren; developed; developed, open space; cultivated crops; water; and wetlands account for about 4 percent of the land cover. Approximately 49 percent of the land in Routt County is publicly owned (Montgomery Watson Harza, 2002). This includes three national forests (Arapaho, Routt, and White River), Bureau of Land Management lands, and lands held by agencies of the state of Colorado (State Land Board, Division of Wildlife, and State Parks). Vegetation is diverse throughout the watershed. Alpine areas are predominantly evergreen and aspen forest, whereas areas around Steamboat Springs

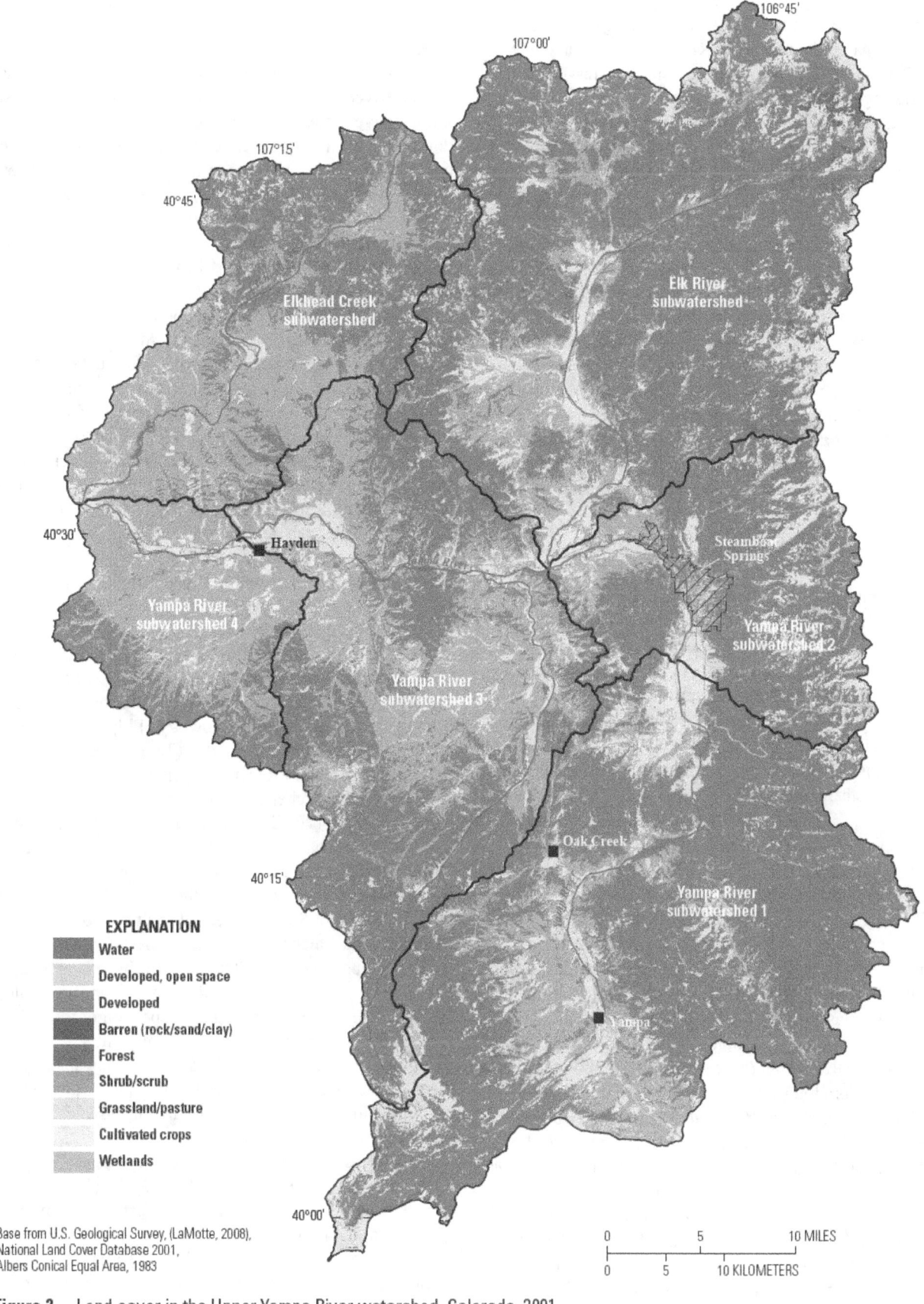

Figure 2.   Land cover in the Upper Yampa River watershed, Colorado, 2001.

are subalpine and generally have Douglas fir, ponderosa pine, and aspen. Areas south from Steamboat Springs, around the town of Yampa, in the lower Elk River valley, and in the lower Yampa River valley are semiarid with shrubs, grassland, and rangeland (fig. 2), which are grazed.

Variation of temperature and precipitation in the UYRW is typical of that found in mountainous and semi-arid regions of Colorado. Annual temperatures in the towns of Steamboat Springs and Hayden range from an average minimum temperature of 0.9 and 4.7° Fahrenheit (F), respectively, during January to an average maximum temperature of 82.6 and 85.6 °F, respectively, during July (High Plains Regional Climate Center, 2010). Frost-free days typically occur during late June through early September in Steamboat Springs and early June through mid-September in Hayden (*http://www.cmg.colostate.edu/gardennotes/749.pdf*, accessed June 2012). Almost 24 inches per year (in/yr) of precipitation falls on average in Steamboat Springs and 17 in/yr falls in Hayden. Much of the precipitation falls as snow throughout the winter months, which melts during spring. Snowfall averages 166 in/yr in Steamboat Springs and 107 in/yr in Hayden.

Hydrology and Water Resources

The Yampa River originates in the Flat Top Mountains as the Bear River, flows northward to the town of Yampa, and becomes the Yampa River in town where Chimney Creek converges with the Bear River (fig. 1). Major tributaries to the Yampa River include Oak Creek, upstream from Steamboat Springs; the Elk River, downstream from Steamboat Springs; and Elkhead Creek, downstream from Hayden. Minor tributaries include Fish Creek east of Steamboat Springs, Fish Creek southwest of Steamboat Springs, Trout Creek, Foidel Creek, and Sage Creek. Streams in the Mount Zirkel, Flat Tops, and Sarvis Creek Wilderness Areas have been classified as outstanding waters by the CDPHE (Colorado Department of Public Health and Environment, 2009a, 2010).

The USGS and the Colorado Division of Water Resources currently ([1]water year 2010) operate 13 streamgage stations in the UYRW (fig. 3, table 1). Streamflow data have been collected during different periods of record at the 13 sites, beginning as early as 1904. Real-time streamflow data for the USGS stations are available at *http://co.water.usgs.gov/infodata/surfacewater.html* (accessed June 2012). Real-time data for the Colorado Division of Water Resources stations are available at *http://www.dwr.state.co.us/SurfaceWater/data/division.aspx?div=6* (accessed June 2012). The USGS also collected streamflow data at 36 additional stations for various periods of record between 1913 and 2008.

Streamflow in the UYRW is dominated by snowmelt, with increasing flows in April, maximum flows in May and June, and decreasing flows in July. Streamflow from August through March is often dominated by base flow from

groundwater discharge. Mean monthly streamflow for 2005 through 2008 for two sites on the Yampa River and one site on Elk River show the seasonal pattern of streamflow in the watershed (fig. 4). Mean monthly streamflow was lowest at Yampa River at Steamboat Springs (site 153) and highest at Yampa River above Elkhead Creek (site 146). Low streamflow during August at Yampa River at Steamboat Springs (site 153) could be of concern because of the possible effects of low streamflow on fish (higher water temperature and lower dissolved oxygen) and on river rafting.

Development of water resources in the UYRW has focused on irrigation, municipal and industrial diversions, and state-sponsored reservoir development (Colorado Water Conservation Board, 2009). Irrigation has changed little since the late 1800s. There have been only small increases in the amount of acreage irrigated since the construction of new ditches and storage systems. The first substantial municipal water system in the UYRW was developed during the 1950s at Steamboat Springs. The City of Steamboat Springs and the Mt. Werner Water and Sanitation District divert most of their municipal water supplies directly from Fish Creek east of Steamboat Springs. When flow in the creek is insufficient for supply, water can be released from Fish Creek Reservoir for augmentation (Colorado Water Conservation Board, 2009). The two public suppliers also can withdraw water from alluvial wells adjacent to the Yampa River. The wells, however, are not a preferred source of municipal water because the quality of the water is not as good as that of the surface-water supplies. Surface water is the primary water source for the Towns of Hayden and Oak Creek. Groundwater is the primary water source for the Towns of Phippsburg and Yampa and part of the source for Hayden (Topper and others, 2003; U.S. Environmental Protection Agency, 2010b).

Two small ditches (maximum water right of 43 cubic feet per second, ft[3]/s) divert water from the UYRW into the Colorado River basin (Colorado Water Conservation Board, 2009). One ditch diverts water from the outlet works of Yamcolo Reservoir (fig. 1), and the other diverts water from the headwaters of Service Creek. The Steamboat Springs Ski Resort diverts water from an alluvial well near the Yampa River just upstream from Steamboat Springs for making artificial snow, typically from October through January (Colorado Water Conservation Board, 2009). A portion of the artificial snowpack is consumptively used during winter and spring; the remaining portion returns to the stream during spring snowmelt. Concerns about the water supply in the watershed are growing with the possibility of future large-scale diversions and increased water demands.

Eight reservoirs in the UYRW can each store 4,000 acre-feet of water or more (table 2). In addition to storage, the water is used for irrigation, recreation, and municipal and industrial purposes. The largest reservoirs are Stagecoach Reservoir, Steamboat Lake, and Elkhead Reservoir (table 2). Because reservoirs in the watershed are small compared to other reservoirs in the Upper Colorado River basin and are located primarily in the headwaters of the Yampa River, streamflow in the Yampa River is largely free-flowing.

---

[1]A water year is the 12-month period from October 1 to September 30. It is designated by the year in which it ends.

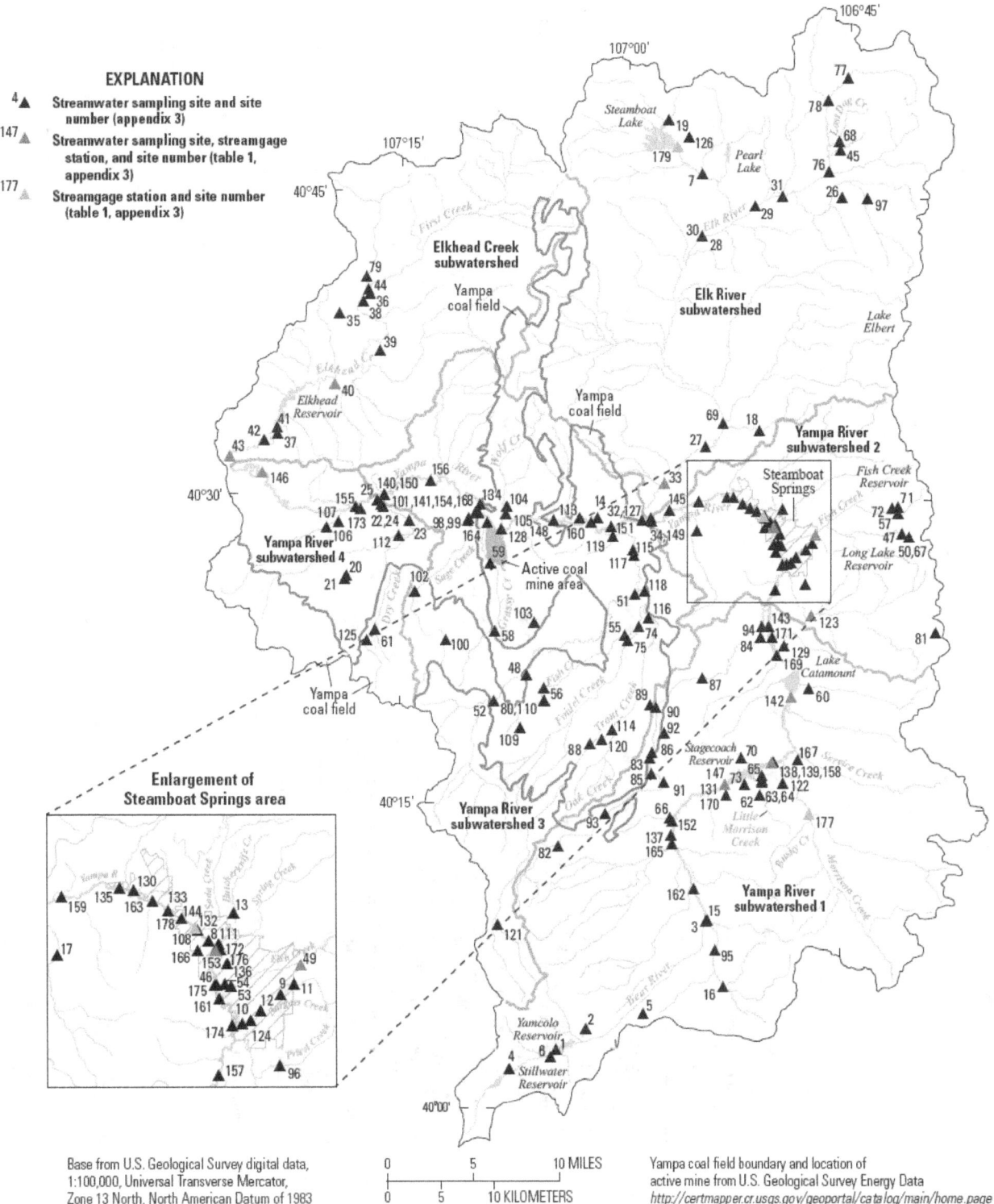

Figure 3. Location of selected stream-water sampling sites and active, water year 2010, streamgage stations, Upper Yampa River watershed, Colorado.

Table 1. Summary of active, water year 2010, streamgage stations in the Upper Yampa River watershed, Colorado.

[USGS, U.S. Geological Survey; CODWR, Colorado Division of Water Resources; WY, water year. USGS information from the USGS National Water Information System (NWIS), *http //waterdata.usgs.gov/nwis.* CODWR information from *http //www.dwr.state.co.us/SurfaceWater/data/division.aspx?div=6.* A water year is defined as a 12-month period beginning October 1 and ending September 30 of the following year. A water year is designated as the year in which it ends.]

Site number (See figure 3)	USGS or CODWR streamgage station identifier	USGS or CODWR streamgage station name	Agency collecting data	Period of streamflow record
177	MORBSCCO	Morrison Creek below Silver Creek	CODWR	4/22/2009–WY 2010, seasonal[1]
147	09237450	Yampa River above Stagecoach Reservoir, CO	USGS	10/1/1988–WY 2010
158	09237500	Yampa River below Stagecoach Reservoir, CO	USGS	10/1/1939–9/30/1944, 10/1/1956–9/30/1972, 10/1/1984–WY 2010
142	YAMABVCO[2]	Yampa R above Lake Catamount nr Steamboat Springs	CODWR	10/1/2003–WY 2010
49	09238900	Fish Cr at Upper Sta nr Steamboat Springs, CO	USGS	10/1/1966–9/30/1972, 5/1/1982–WY 2010
123	WLTNCKCO[3]	Walton Creek near Steamboat Springs, CO.	CODWR	10/1/1965–WY 2010
153	09239500	Yampa River at Steamboat Springs, CO	USGS	10/1/1904–9/30/1906, 3/1/1910–WY 2010
178	09240020	Yampa River below Soda Creek at Steamboat Spgs, CO	USGS	6/25/2008–WY 2010, seasonal[4]
179	WILBSLCO	Willow Creek below Steamboat Lake	CODWR	10/1/1978–WY 2010
33	09242500	Elk River near Milner, CO	USGS	10/1/1904–9/30/1906, 10/1/1909–9/30/1927, 4/1/1990–WY 2010
146	09244490	Yampa River above Elkhead Creek near Hayden, CO	USGS	3/16/2004–WY 2010
40	09246200	Elkhead Creek above Long Gulch, near Hayden, CO	USGS	8/10/1995–WY 2010
43	09246500	Elkhead Creek near Craig, CO	USGS	1/1/1910–12/31/1918, 7/11/2008–WY 2010

[1]Site is operated seasonally from late April through October.

[2]Site is located at or near Colorado Department of Public Health and Environment site 12807.

[3]Site also is USGS site 09238500.

[4]Site is operated seasonally from May through August.

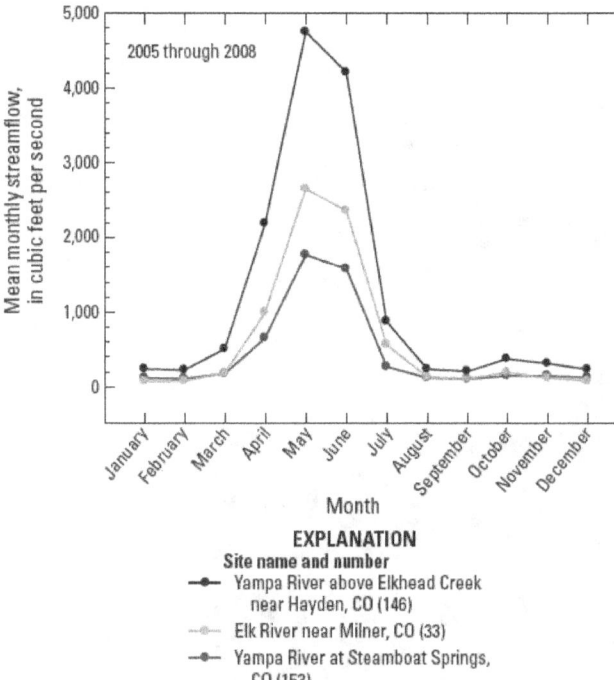

Figure 4. Mean monthly streamflow for 2005 through 2008 for selected streamgage stations in the Upper Yampa River watershed, Colorado.

Table 2. Reservoirs in the Upper Yampa River watershed, Colorado, with active storage capacity of 4,000 acre-feet or more.

[ac-ft, acre feet. Multiple purpose means irrigation, recreation, municipal, and industrial uses. Data from Colorado Water Conservation Board (2009)]

Reservoir (see figure 1)	Active storage capacity (ac-ft)	Purpose[1]
Stillwater Reservoir	5,175	Irrigation
Fish Creek Reservoir	4,042	Municipal
Pearl Lake	5,657	Fishery and recreation
Steamboat Lake	23,064	Fishery and recreation
Elkhead Reservoir	10,422	Multiple
Lake Catamount	7,422	Fishery and recreation
Yamcolo Reservoir	8,028	Irrigation
Stagecoach Reservoir	30,000	Multiple

[1]In addition to storage.

Geology

The UYRW is underlain by rocks of Precambrian age to unconsolidated Quaternary alluvium (fig. 5) (Tweto, 1979). The oldest rocks are in the eastern one-third (western side of Gore and Parks Ranges) of the watershed. These mountainous areas are underlain by igneous (granitic and mafic) rocks and metamorphic (gneiss, schist, and migmatite) rocks. Permian and Triassic sedimentary rock (sandstone, shale, siltstone) occupy a small portion of the area north from Steamboat Springs (fig. 5). The western two-thirds of the watershed is underlain by sedimentary rocks of Cretaceous age and sedimentary and igneous rocks of Tertiary age. Dominant Cretaceous rock types include

sandstones and shale, and major coals beds composing the UYRW portion of the Yampa coal field (fig. 1). Tertiary rocks include sandstones and shales. Broad valleys and small rounded hills are present in areas with less resistant shales, and ridges and mesas are present in areas with more resistant sandstones. The Tertiary basalts and intrusive and volcanic rocks form the Flat Top Mountains in the southern portion of the watershed and the area north from Steamboat Lake. The youngest formations include the unconsolidated surficial deposits and rocks of Quaternary age, including landslide and valley-fill deposits, glacial drift, gravels, and alluvium. Landslide deposits are most common in the Flat Tops Mountains. Glacial drift, primarily from Pinedale and Bull Lake glaciations, occurs on the western side of the Gore and Park Ranges. The youngest gravels and alluvium are located primarily along the Yampa and Elk Rivers.

Cretaceous and Tertiary sedimentary rocks contain soluble materials and trace elements, including arsenic, barium, boron, iron, manganese, nickel, selenium, and strontium. Formations that are seleniferous include the Lewis Shale, Williams Fork Formation and Iles Formation of the Mesaverde Group, and Mancos Shale (Butler and others, 1996; Stephens and Waddell, 1998). Major coal deposits in the Yampa coal field occur in the Upper Cretaceous Williams Fork and Iles Formations in northwestern Colorado (Johnson and others, 2000). In the eastern part of the coal field that is in the UYRW, the principal coal beds in the middle coal group of the Williams Fork Formation are, in ascending order, Wolf Creek, Wadge, and Lennox (Bass and others, 1955). The coal deposits were formed in alternating mixed marine and nonmarine environments at the western edge of the Late Cretaceous Western Interior Seaway. Most sulfur in the deposits is in the form of organic sulfur rather than pyritic sulfur and sulfate sulfur (Affolter, 2000). High contents of strontium, barium, and phosphorus are found in the Iles Formation compared to other Cretaceous coals in the Colorado Plateau (Arizona, Colorado, New Mexico, and Utah) (Affolter, 2000). Similarly, high contents of arsenic and manganese are found in the Lennox and Wolf Creek coal beds of the Williams Fork Formation. Currently (2012), one underground coal mining operation, at the Twentymile Mine, is active in the UYRW portion of the Yampa coal field (fig. 1).

More than 150 hot springs are located in the Steamboat Springs area (Frazier, 2000). Thermal waters for the springs probably are meteoric waters that have been heated at depths of 12,000 to 15,000 ft (Lund, 2006). Heated water most likely rises to the surface from a network of faults and fractures that cross the region. Chemical constituents in the springs include sodium, chloride, sulfate, bicarbonate, and lithium.

Methods of Data Compilation, Review, and Analysis

Various methods of data compilation, review, and analysis of water-quality data were used to assess water-quality conditions in the UYRW. Data from several sources were compiled into a web-accessible UYRW water-quality database.

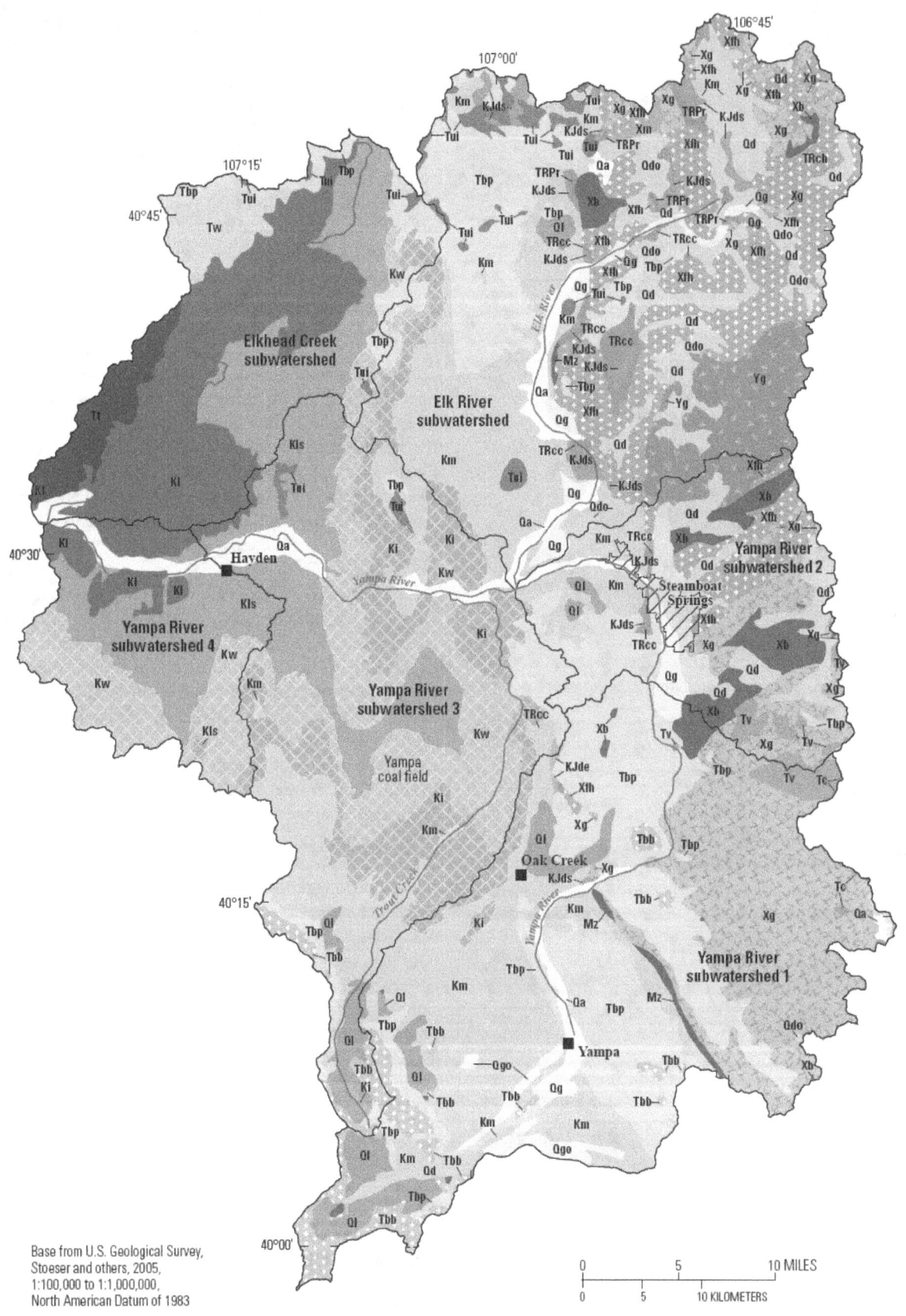

Base from U.S. Geological Survey,
Stoeser and others, 2005,
1:100,000 to 1:1,000,000,
North American Datum of 1983

EXPLANATION

Qa	Quaternary modern alluvium
Qg	Quaternary gravels and alluvium
Qgo	Quaternary older gravels and alluviums
Qd	Quaternary glacial drift of Pinedale and Bull Lake Glaciations
Qdo	Quaternary older glacial drifts
Ql	Quaternary landslide deposits
Tbp	Tertiary Browns Park Formation
Tw	Tertiary Wasatch Formation
Tc	Tertiary Coalmont Formation
Tf	Tertiary Fort Union Formation
Tbb	Tertiary basalt flows and tuff, breccia, and conglomerate of late-volcanic bimodal suite
Tv	Tertiary volcanic rocks
Tui	Upper Tertiary intrusive rocks of 20 million years before present (Ma)
Kl	Cretaceous Laramie Formation-shales, claystone, sandstone, and major coal beds
Kls	Cretaceous Lewis Shale
Kw	Cretaceous Williams Fork Formation-sandstone, shale, and major coal beds
Kd	Cretaceous Dakota Sandstone
Ki	Cretaceous Iles Formation-sandstone and shale.
Km	Cretaceous Mancos Shale
KJde	Cretaceous and Jurassic Dakota, Burro Canyon, Morrison, Wanakah, and Entrada formations
KJds	Cretaceous and Jurassic Dakota, Morrison, and Sundance formations
Jms	Jurassic Morrison Formation and Sundance Formation (shale and siltstone), and Entrada Sandstone
TRch	Triassic Chugwater Formation
TRcc	Triassic Chinle and Chugwaters Formations
TRPr	Triassic and Permian rocks
Mz	Mesozoic rocks
Xb	Precambrian bioitite gneiss, schist, and migmatite
Xfh	Precambrian felsic and hornblende gneisses
Yg	Precambrian granitic rocks of 1,400 Ma
Xg	Precambrian granitic rocks of 1,700 Ma
Xm	Precambrian mafic rocks of 1,700 Ma
	Water
	Yampa coal field

Ma   Mega Annum (millions of years before present)

Figure 5 (above and facing). Geology of the Upper Yampa River watershed, Colorado.

Quality-assurance checks and data aggregation steps were applied to pertinent data retrieved from the online database. The resulting physical-property and chemical constituent data were summarized statistically, displayed spatially in figures, analyzed, and interpreted. Data were analyzed for temporal trends and compared to state of Colorado and federal water-quality standards and recommendations. The following discussion describes methods used to process, evaluate, and interpret water-quality data for the UYRW.

Data Sources and Water-Quality Database

Data summarized and analyzed in this study were collected and reported by the U.S. Environmental Protection Agency (USEPA), U.S. Department of Agriculture Forest Service, USGS, Colorado Department of Agriculture (CDOA), Colorado Department of Public Health and Environment (CDPHE), Colorado Division of Wildlife Riverwatch Program, and the City of Steamboat Springs (data collected by GEI Consultants, Inc.) (table 3). Data originally were obtained in an electronic format from the USGS National Water Information System, USEPA STOrage and RETrieval (STORET) Data Warehouse, and GEI Consultants, Inc., and were merged to form the UYRW water-quality database (*http://rmgsc.cr.usgs.gov/cwqdr/Yampa/index.shtml*). The database primarily contains data for physical properties and chemical constituents in samples collected from surface water (streams, canals, diversions, lakes, and reservoirs), groundwater (wells, springs, and seeps), mining discharges (tunnel, shaft, or mine), and wastewater-treatment-plant discharges, including effluent, during most years from 1944 through 2009. However, the year 1975 was chosen as the starting year for data analysis because major development of coal resources in the watershed began that year (Colorado Geological Survey, 2005). Data for water-quality constituents are reported as filtered (through a 0.45 micron filter, described as "dissolved") or unfiltered. Data can also be reported as total (for example, total nitrogen, which includes all forms of nitrogen) or total recoverable (which involves incomplete digestion of particulate matter in the laboratory).

Data Quality Assurance

After data for selected physical properties and chemical constituents were retrieved from the UYRW water-quality database, a number of quality-assurance procedures were applied to the data. The USEPA has established low and high values for 190 water-quality properties and constituents as an edit-checking procedure for data entered into STORET since November 1993 (National Park Service, 2001). Low and high values for 63 physical properties and chemical constituents were used for edit-checking of water-quality data retrieved from the UYRW database (appendix 1). Using the edit-checking procedures, temperature values of −17.8 and 90 degrees Celsius (°C) and a dissolved oxygen concentration of 779 milligrams per liter (mg/L) were deleted. Trace-element data greater than the high values were not deleted. Trace elements can occur in very high concentrations in areas with historical mining activities or in areas with naturally high mineralization because of geology. For a given individual sample with dissolved- and unfiltered-concentration data, the dissolved concentration was checked against the unfiltered concentration. Both data values would be marked for deletion if the dissolved

Table 3. Number of sample sites, period of water-quality record, and number of water-quality and biological samples collected by various organizations at selected site types in the Upper Yampa River watershed, Colorado, 1975 through 2009.

[No., number; --, no data]

Organization collecting samples	Streams			Lakes and reservoirs			Groundwater wells			Macroinvertebrates		
	No. of sites	Period of water-quality record	No. of samples	No. of sites	Period of water-quality record	No. of sample days[1]	No. of sites	Period of water-quality record	No. of samples	No. of sites	Period of water-quality record	No. of sample days
City of Steamboat Springs (GEI Consultants, Inc.)	8	2007–2008	32	--	--	--	--	--	--	4	2005, 2007–2008	[2]12
Colorado Department of Agriculture[3]	--	--	--	--	--	--	6	1998	6	--	--	--
Colorado Department of Public Health and Environment[3]	56	1975–2007	866	[4]4	1990, 2006	4	--	--	--	37	[5]1997–2008	61
Colorado Division of Wildlife Riverwatch Program[3]	14	1990–2004	1,046	--	--	--	--	--	--	--	--	--
U.S. Environmental Protection Agency[3]	2	2000–2001	4	1	1985	1	--	--	--	--	--	--
U.S. Forest Service[3]	3	1975, 1978–79	35	--	--	--	--	--	--	--	--	--
U.S. Geological Survey[6]	128	1975–2009	3,878	37	1975–76, 1978–79, 1983, 1985–2009	364	322	[7]1975–1989	[8]1,574	25	1975	25
Total	211		5,861	42		369	328		1,580	66		98

[1]Samples collected at multiple depths on the same day are counted as one sample day.

[2]Four samples were collected on each sample day at a site.

[3]Data retrieved from U.S. Environmental Protection Agency's STOrage and RETrieval (STORET) Data Warehouse.

[4]Two sites in Stagecoach Reservoir are at the same location and are counted as one site, as are two sites in Steamboat Lake. For each body of water, samples collected near the top and bottom of the water column at the same location have different site identification numbers in the Upper Yampa River watershed database but are counted as one site in this table.

[5]No data for 1999, 2002, 2004, and 2007.

[6]Data retrieved from the National Water Information System (NWIS).

[7]No data for 1985.

[8]Count consists of 810 samples with water-quality data and 764 samples with water-level data only. Count of water-quality samples does not include multiple (14) water-quality samples collected on one day from selected wells.

concentration was greater than the unfiltered concentration by more than 10 percent. All dissolved and unfiltered data retrieved from the UYRW database met this edit-check procedure. An ion charge balance was calculated for individual samples with sufficient data (calcium, magnesium, sodium, potassium, alkalinity, chloride, and sulfate concentrations) to calculate the balance. All charge balances were within 10 percent, and no data were deleted. One dissolved iron concentration for a particular site was 100-times greater than the next highest concentration at the site and was excluded from this study. Twelve measurements of instantaneous discharge with a value of 0 were deleted.

Many of the reported concentrations for water-quality constituents showed concentrations of 0. According to Hem (1992), concentrations of 0 can imply that the amount of a constituent present in a sample was less than a detection level and that the analytical procedure could not detect a concentration less than the detection level. For 427 individual analyses of stream-water samples, constituent concentrations with a value of 0 were converted to the minimum detection level for the constituent of interest. Most (311 of 427) reported concentrations of 0 in stream-water samples were for trace elements, but 0 was also reported for nutrients (108), major ions (5), and dissolved uranium (3). Constituent concentrations for 438 individual analyses of groundwater samples were converted from a 0 value to the lowest detection level; values of 0 were reported for trace elements (353) or nutrients (85). A total of 801 individual analyses of stream-water samples and 620 individual analyses of groundwater samples resulted in a data remark code of M (presence of constituent verified but not quantified); these data were excluded from the analysis for this study.

For a large portion of the water-quality data, limited metadata and (or) quality-assurance data were available. Therefore, it is possible that some data may contain errors that were not detected during the quality-assurance review. Assumptions regarding water-quality collection methods and laboratory analytical techniques used by different data sources were made based on available information. No distinctions between water-quality data collection methods and laboratory analytical techniques were made when metadata were unavailable to support these distinctions. Disparities between data from different sources that resulted from differences in water-quality collection methods and laboratory analytical techniques could affect the precision and accuracy of the statistical results. Although the effect of methodological differences could not be quantified in this analysis, robust statistical methods were used to limit the influence of outliers on statistical results of the analysis. It has been documented for USGS trace-element data that dissolved concentrations of arsenic, boron, beryllium, cadmium, chromium, copper, lead, mercury, and zinc collected before 1992 may have been contaminated during sample collection and field processing (USGS, 1991). An in-depth review of these data for this study could not be conducted because field quality-assurance

data were not available for analysis. For all data that could not be quality assured, it is assumed that measurements made at the time of sampling and results from laboratory analyses are of good quality. These data are used as reported in the UYRW database.

Data Aggregation

Data on physical properties and chemical constituents were compiled from the various sources, each with differing laboratory methods and sampling and reporting conventions. For many physical properties and chemical constituents, data were available under one or more parameter codes for the same property or constituent. Equivalent parameter codes for different physical properties and chemical constituent groups were combined for data-analysis purposes by using the data aggregation groups summarized in table 4. Measurement or concentration data for the first parameter code listed in the table for each constituent were preferred for analysis. If these data were not available, data for the second parameter code listed were used. This aggregation continued for each parameter code in a constituent group, resulting in a single constituent name with one reporting convention. All of the aggregated nitrate data, for example, are in the form of nitrate as nitrogen instead of nitrate as nitrate. Procedures used to aggregate nutrient data follow those used by Mueller and others (1995).

In natural waters, ammonia nitrogen can be in the form of aqueous ammonia (un-ionized ammonia, NH_3) or ammonium (NH_4^+). In most natural waters (pH less than 9.24), ammonia nitrogen occurs as ammonium rather than un-ionized ammonia (Hem, 1992). For the USGS and other federal and state agencies, the sum of un-ionized ammonia concentrations and ammonium ions is reported as "ammonia" or "total ammonia." In this report, the sum is reported as "total ammonia."

Censored Values

Computing summary statistics for the water-quality data presented in this report were complicated by the presence of multiple detection or reporting levels for censored data for many chemical constituents. Censored data for this study are data reported as "less than" a particular laboratory detection level or reporting level. When water-quality results were reported with censored (below laboratory detection or reporting level) values, estimates of percentile values, including the 50th percentile or median, were calculated using the Kaplan-Meier, adjusted maximum likelihood estimation (AMLE), or regression of ordered statistics (ROS) methods following the recommendations of Helsel (2005). The Kaplan-Meier method was used when less than 50 percent of the data were censored values. When censored values were 50 to 80 percent, the ROS method was used when there were fewer than 50 samples or when there were 50 or more samples and the distribution of data was not normal.

Table 4. Summary of procedures used to aggregate data for selected physical properties, streamflow, or water-quality constituents for streams, Upper Yampa River watershed, Colorado

[No., number; μS/cm, microsiemens per centimeter at 25 degrees Celsius; C, degrees Celsius; mg/L, milligrams per liter; CaCO$_3$, calcium carbonate; Ca, calcium; Mg, magnesium; ft^3/s, cubic feet per second; S, sulfur; N, nitrogen; NO$_2$ nitrite; *, constituent computed from procedure listed for nitrite, as N; NO$_3$, nitrate; NH$_3$, un-ionized ammonia; NH$_4^+$, ammonium; P, phosphorus; PO$_4$, phosphate; μg/L, micrograms per liter; Fe, iron; col/100 mL, colonies per 100 milliliters; m-TEC MF, membrane-Thermotolerant *Escherichia coli* Membrane Filtration; M-FC MF, membrane-Fecal Coliform Membrane Filtration. Procedures used to aggregate nutrient data follow those used by Mueller and others (1995)]

Physical property or constituent (reporting units)	Physical property or constituent name	Parameter code[1]	No. of samples
Physical properties and streamflow			
Specific conductance (μS/cm)	Specific conductance, water, unfiltered, μS/cm at 25 °C	00095	3,890
	Specific conductance, water, unfiltered, field, μS/cm at 25 °C	00094	415
pH (standard units)	pH, water, unfiltered, field, standard units	00400	3,104
	pH, water, unfiltered, laboratory, standard units	00403	36
Hardness (mg/L as CaCO$_3$)	Hardness, water, mg/L as CaCO$_3$	00900	1,615
	Hardness, water, total, Ca Mg calculated, mg/L as CaCO$_3$	46570	787
	Hardness, water, filtered, calculated, mg/L as CaCO$_3$	00906	32
Acid neutralizing capacity (mg/L as CaCO$_3$)	Acid neutralizing capacity, water, unfiltered, fixed endpoint (pH 4.5) titration, field, mg/L as CaCO$_3$	00410	476
	Acid neutralizing capacity, water, unfiltered, fixed endpoint (pH 4.5) titration, laboratory, mg/L as CaCO$_3$	90410	716
	Acid neutralizing capacity, water, unfiltered, mg/L as CaCO$_3$	00431	1,052
Streamflow (ft^3/s)	Discharge, instantaneous, ft^3/s	00061	3,665
	Discharge, ft^3/s	00060	18
	Flow, estimated, stream, ft^3/s	74069	1
Dissolved solids			
Dissolved solids, (mg/L)	Residue, water, filtered, sum of constituents, mg/L	70301	989
	Residue on evaporation, dried at 180 °C, water, filtered, mg/L	70300	754
Major ions			
Bicarbonate, dissolved (mg/L)	Bicarbonate, water, unfiltered, fixed endpoint (pH 4.5) titration, field, mg/L	P00440	160
	Bicarbonate, water, filtered, inflection-point titration method (incremental titration method), field, mg/L	P00453	24
Sulfate, dissolved (mg/L)	Sulfate, water, filtered, mg/L	00945	1,018
	Sulfate, water, dissolved, as S, mg/L	78462	2
	(multiplied by 3.0)		
Chloride, dissolved (mg/L)	Chloride, water, filtered, mg/L	00940	1,014
	Chloride, dissolved in water, mg/L	00941	2
Nutrients			
Nitrite, unfiltered (mg/L as N)	Nitrite, water, unfiltered, mg/L as N	00615	34
	Nitrite, water, unfiltered, mg/L as NO$_2$	71855	61
	(multiplied by 0.30446)		
Nitrate, dissolved (mg/L as N)	Nitrate plus nitrite, water, filtered, mg/L as N, minus [2]nitrite, filtered, mg/L as N	00631	1,092
		00613	628
	Nitrate, water, filtered, mg/L as N	00618	4
Nitrate, unfiltered (mg/L as N)	Nitrate plus nitrite, water, unfiltered, mg/L as N, minus [2]nitrite, unfiltered, mg/L as N	00630	1,155
		*	95
	Nitrate, water, unfiltered, mg/L as NO$_3$	71850	62
	(multiplied by 0.2259)		

Table 4. Summary of procedures used to aggregate data for selected physical properties, streamflow, or water-quality constituents for streams, Upper Yampa River watershed, Colorado.—Continued

[No., number; µS/cm, microsiemens per centimeter at 25 degrees Celsius; C, degrees Celsius; mg/L, milligrams per liter; CaCO₃, calcium carbonate; Ca, calcium; Mg, magnesium; ft³/s, cubic feet per second; S, sulfur; N, nitrogen; NO₂, nitrite; *, constituent computed from procedure listed for nitrite, as N; NO₃, nitrate; NH₃, un-ionized ammonia; NH₄⁺, ammonium; P, phosphorus; PO₄ phosphate; µg/L; micrograms per liter; Fe, iron; col/100 mL; colonies per 100 milliliters; m-TEC MF, membrane-Thermotolerant *Escherichia coli* Membrane Filtration; M-FC MF, Membrane-Fecal Coliform Membrane Filtration. Procedures used to aggregate nutrient data follow those used by Mueller and others (1995)]

Physical property or constituent (reporting units)	Physical property or constituent name	Parameter code[1]	No. of samples
Nutrients—Continued			
Total ammonia, unfiltered (mg/L as N)	Ammonia, water, unfiltered, mg/L as N	00610	184
	Ammonia + ammonium as N, total $NH_3 + NH_4^+$ in water, mg/L	82230	843
Orthophosphate (mg/L as P)	Orthophosphate, water, filtered, mg/L as P	00671	836
	Orthophosphate, water, unfiltered, mg/L as P	70507	16
	Phosphate, water, unfiltered, mg/L as PO_4	00650	2
	(multiplied by 0.3261)		
Trace elements			
Aluminum, total recoverable (µg/L)	Aluminum, water, unfiltered, recoverable, µg/L	01104	51
	Aluminum, water, unfiltered, recoverable, µg/L	01105	304
Chromium, dissolved (µg/L)	Chromium (VI), water, filtered, µg/L	01032	21
	Chromium, water, filtered, µg/L	01030	133
Chromium, total recoverable (µg/L)	Chromium, water, unfiltered, recoverable, µg/L	01034	172
	Chromium, water, unfiltered, recoverable, µg/L	01118	59
Copper, total recoverable (µg/L)	Copper, water, unfiltered, recoverable, µg/L	01042	724
	Copper, water, unfiltered, recoverable, µg/L	01119	253
Iron, total recoverable (µg/L)	Iron, total recoverable in water as Fe µg/L	00980	690
	Iron, water, unfiltered, recoverable, µg/L	01045	1,339
Lead, total recoverable (µg/L)	Lead, water, unfiltered, recoverable, µg/L	01051	281
	Lead, water, unfiltered, recoverable, µg/L	01114	264
Manganese, total recoverable (µg/L)	Manganese, water, unfiltered, recoverable, µg/L	01055	1,228
	Manganese, water, unfiltered, recoverable, µg/L	01123	268
Nickel, total recoverable (µg/L)	Nickel, water, unfiltered, recoverable, µg/L	01067	186
	Nickel, water, unfiltered, recoverable, µg/L	01074	26
Silver, total recoverable (µg/L)	Silver, water, unfiltered, recoverable, µg/L	01077	88
	Silver, water, unfiltered, recoverable, µg/L	01079	84
Zinc, total recoverable (µg/L)	Zinc, water, unfiltered, recoverable, µg/L	01092	650
	Zinc, water, unfiltered, recoverable, µg/L	01094	32
Coliform bacteria			
Escherichia coli (col/100 mL)	*Escherichia coli*, modified m-TEC MF method, water, col/100 mL	90902	21
	Escherichia coli, m-TEC MF method, water, col/100 mL	31633	101
Fecal coliform (col/100 mL)	Fecal coliform, M-FC MF (0.7 micron) method, water, col/100 mL	31613	127
	Fecal coliform, M-FC MF (0.7 micron) method, water, col/100 mL	31625	153
	Fecal coliform, M-FC MF (0.45 micron) method, water, col/100 mL	31616	119

[1]Parameter codes are from the U.S. Geological Survey National Water Information System (NWIS) and the U.S. Environmental Protection Agency Data STOrage and RETrieval (STORET) Data Warehouse.

[2]Values less than the detection level and missing values were not included in the calculation.

For 50 or more samples and a normal distribution, the AMLE method was used. When 80 percent or more of the data were censored values, the median value was not computed; only the minimum and maximum values are reported in summary tables. Data from 11 analyses with a concentration reported as "greater-than" a particular concentration were excluded from this study. Some water-quality results that are less than laboratory detection levels were reported as "estimated" (E) values rather than censored values. Because of improvements in laboratory analytical techniques, a chemist is able to report an estimated concentration when a compound meets all criteria for identification, but the concentration value is less than the laboratory reporting level (Childress and others, 1999).

Water-Quality Standards

In-stream water-quality standards for surface water in the state of Colorado have been established by the CDPHE to protect the beneficial uses of surface water, including protection of cold- and warm-water aquatic life, recreation, water supply, and agriculture (Colorado Department of Public Health and Environment, 2009a, 2010). The standards are applied to stream segments and water bodies on the basis of water-use classification (appendix 2) (Colorado Department of Public Health and Environment, 2009a, 2010). Water-quality standards can vary between segments of different streams and between different segments on the same stream. For this study, the authors assigned a stream segment to each stream site on the basis of the segment descriptions in the water-quality standards table for the Upper Colorado River basin (Colorado Department of Public Health and Environment, 2010). In-stream water-quality standards have not been established for all physical properties and chemical constituents.

Two types of numeric water-quality standards, fixed values and table values standards (TVSs), have been established by the CDPHE. Fixed values are maximum contaminant levels (MCLs), secondary maximum contaminant levels (SMCLs), and a recreation-based standard (appendix 2). The MCLs and SMCLs apply to drinking water from public water systems. An MCL is a legally enforceable standard for drinking water; an SMCL is a nonenforceable guideline for contaminants that may have cosmetic (skin or tooth discoloration), aesthetic (taste, odor, and color), or technical (corrosion and staining) effects. For streams, lakes, or reservoirs in the UYRW that are not drinking-water supplies, the MCLs are used in this study as guidelines for interpreting water quality. Table value standards, established for aquatic-life protection, are calculated values based on published formulas and include acute and chronic classifications. Except for temperature, an acute standard is a value that is not to be exceeded by a constituent concentration in a single sample or by an average concentration for all samples collected during a one-day period. A chronic standard is a value not to be exceeded by a concentration in a single representative sample or by an average concentration for all samples collected during a 30-day period. Two types of TVSs have

been established by CDPHE for water temperature. The Daily Maximum Temperature (DM) is the highest 2-hour average water temperature recorded for a given 24-hour period, and the Maximum Weekly Average Temperature (MWAT) is the maximum average of multiple, equally spaced daily water temperature measurements collected over 7 consecutive days (Colorado Department of Public Health and Environment, 2009a). The TVS for total ammonia varies depending on fish species, pH, and water temperature. Table value standards for trace elements vary depending on stream hardness. Hardness-dependent TVSs were calculated using the mean of available hardness data that corresponds to the trace-element data. If these hardness data were not available, the median hardness value for the site, other sites on the same stream, or sites on nearby streams was used.

The CDPHE has not established an in-stream water-quality standard for total phosphorous. The USEPA, however, has recommended that total phosphorous concentrations be less than 0.1 mg/L for streams that do not flow directly into lakes and reservoirs and less than 0.05 mg/L for streams that do flow directly into lakes and reservoirs to control eutrophication of the water bodies (U.S. Environmental Protection Agency, 2000). Unfiltered total phosphorus data for the UYRW are compared to the USEPA recommended concentrations to provide an environmentally relevant context to concentrations of total phosphorus in steams. Streams used in the comparison are those with standards for other nutrients.

For this study, determination of attainment or nonattainment of a standard for a stream site is based on the 15th and 85th percentiles of the data for pH, the 15th percentile of the data for dissolved oxygen, the 85th percentile of data for dissolved and unfiltered constituents, and the 50th percentile of the data for total recoverable constituents (Water Quality Control Division, 2002) in samples with more than two data points. As required under Section 303(d) of the federal Clean Water Act, the CDPHE established the 303(d) list of impaired waters and monitoring and evaluation list (table 5) (Colorado Department of Public Health and Environment, 2012).

Water-quality standards for groundwater in the state of Colorado have been established by the CDPHE to protect the beneficial uses of groundwater (Colorado Department of Public Health and Environment, 2009b). Water-quality data for groundwater in the UYRW are compared to standards for agricultural use in livestock watering and "domestic uses," which are existing or potential future uses of groundwater for household or family uses, including drinking, gardening, municipal, and (or) farm uses (Colorado Department of Public Health and Environment, 2009b). Standards for domestic uses include human-health (HH) standards and SMCLs. HH standards have been established to protect the public from acute and chronic effects from exposure to contaminated water (Colorado Department of Public Health and Environment, 2009b). Many HH standards are MCLs. Comparison of water-quality data to groundwater standards is used as a guideline for interpreting water quality but is not used for legally enforceable purposes.

Table 5.   Colorado Department of Public Health and Environment Section 303(d) list of impaired waters and monitoring and evaluation list for the Upper Yampa River watershed, Colorado, 2012.

[--, no listing; Mn, manganese; Se, selenium; Hg, mercury; FCA, fish consumption advisory; USFS, U.S. Department of Agriculture Forest Service; Fe, iron; dis, dissolved; Zn, zinc; DO, dissolved oxygen; *E. coli*, *Escherichia coli*; trec, total recoverable; Pb, lead. Data from Colorado Department of Public Health and Environment (2012). Description of stream segments are in Colorado Department of Public Health and Environment (2010)]

Stream segment	Segment description	Segment portion	Clean Water Act Section 303(d) impairment	Monitoring and evaluation list parameter
2a	Mainstem of the Yampa River from Wheeler Creek to Oak Creek	Yampa River below Stagecoach Reservoir	--	Mn, Se
2b	All lakes and reservoirs tributary to the Yampa River, Elkhead Creek, and the Little Snake River	Elkhead Reservoir, Lake Catamount	Aquatic life use (Hg FCA)	--
2c	Yampa River, from Oak Creek to Elkhead Creek	All	--	Water temperature
3	All tributaries to Yampa River, except for specific listings on USFS land	Bushy Creek	Sediment	--
3	All tributaries to Yampa River, except for specific listings on USFS land	Walton Creek	--	Mn
3	All tributaries to Yampa River, except for specific listings on USFS land	Little Morrison Creek	--	Fe (dis), Zn
4	Little White Snake Creek, source to Yampa River	All	--	DO
8	Elk River source to Yampa River	Elk River below Morin Ditch	*E. coli*	--
8	Elk River including tributaries and wetlands from the source to Yampa River	Lost Dog Creek	--	Hg
13b	Foidel Creek and tributaries, Fish Creek, Middle Creek and tributaries	Fish Creek	--	*E. coli*
13d	Dry Creek including all tributaries and wetlands from the source to the Yampa River	Below Seneca sample location 8 (WSD5)	Se	--
13d	Dry Creek including all tributaries and wetlands from the source to the Yampa River	All	Fe (trec) (snowmelt season)	--
13d	Dry Creek including all tributaries and wetlands from the source to the Yampa River	Dry Creek below Routt County Road 53	--	*E. coli*, Pb
13e	Sage Creek, Grassy Creek and tributaries	Sage Creek below Routt County Road 51D	Se	--
15	Mainstem of Elkhead Creek and tributaries Calf Creek and 80A Road on the Dry Fork of Elkhead Creek, to confluence with the Yampa River	Elkhead Creek	Aquatic life (provisional)[1]	--

[1]Listing is based on macroinvertebrate data rather than water-quality data.

Temporal Trend Analysis

Trend analyses were performed with the TIBCO Spotfire® S+program using the USGS library package ESTREND (Schertz and others, 1991). Analyses were conducted on selected physical properties and chemical constituents for stream sites with at least 10 years of quarterly data for a period of record ending after 2000 and with less than 10-percent censored data. Selected physical properties and chemical constituents were analyzed by using the seasonal Kendall test (Hirsch and others, 1982; Helsel and Hirsch, 2002). The seasonal Kendall test, a nonparametric rank-based procedure, was used to analyze water-quality data for monotonic changes in concentrations with time. Seasonality in water-quality data are accounted for by comparing data for different seasons; for example, data for January through March are compared only with data for January through March, April through June with April through June, and so forth. Because of the strong correlation that exists between many water-quality

constituents and streamflow, most water-quality data were flow-adjusted prior to testing for trends, using the streamflow measured at the time of sample collection. Flow adjustment removes that variability in concentration that is related to natural changes in streamflow, allowing for trends caused by other means, such as human activities, to be more readily identified. When flow-adjustment was applied to the data, the data were regressed against streamflow, and the residuals of the resulting equation were used in the seasonal Kendall test. Most often, the best flow-adjustment procedure is to regress the log-transformed constituent data against log-transformed streamflow (D.K. Mueller, U.S. Geological Survey, oral commun., 2010).

A trend was determined to be present when the p-value of the statistical test was less than 0.05. The smaller the p-value, the stronger the evidence is to reject the null hypothesis that there is no relation between concentration and time: that no trend exists (Helsel and Hirsch, 2002). A trend in an upward direction was identified when a constituent concentration increased more often over time than it decreased. The estimated trend, in percent per year, is reported from the model calculations. The trend slope is an estimate of the yearly change in a value or concentration for the tested time period and is presented as a percentage of the median value per year or concentration per year. Trend results differ depending on the time period used in the trend analysis. Trends that are identified for a particular constituent at a site for one time period may not be identified when another time period is used. A trend also may be statistically significant but not environmentally significant. For example, a statistically significant upward trend with a rate of change of 0.5 mg/L per year for a constituent may have little environmental relevance if the average concentration is 100 mg/L.

Water-Quality Assessment

This section summarizes water-quality data for streams, lakes, reservoirs, and groundwater in the UYRW. General information on physical properties and chemical constituents are presented first, followed by detailed analyses of water-quality data for physical properties, major ions, nutrients, trace elements, uranium, coliform bacteria, and (or) suspended sediment for selected streams, lakes, reservoirs, and groundwater.

Streams

For this study, water-quality data from the UYRW database were compiled for 211 stream sites from January 1975 through September 2009. A total of 176 site locations were unique; some stream sites from different agencies had the same physical location (fig. 3, appendix 3). Site names in appendix 3 are from the UYRW water-quality database; site names used in the body of this report are shorter versions of those in appendix 3. Because streams in the UYRW can have distinct water-quality characteristics based on location in the watershed, the watershed was divided into six subwatersheds for data analysis: Yampa River and tributaries upstream from Chuck Lewis State Wildlife Area (Yampa River subwatershed 1), Yampa River and tributaries from Chuck Lewis State Wildlife Area to Elk River confluence (Yampa River subwatershed 2), Elk River and tributaries (Elk River subwatershed), Yampa River and tributaries from Elk River confluence to Town of Hayden (Yampa River subwatershed 3), Yampa River and tributaries from Town of Hayden to Elkhead Creek confluence (Yampa River subwatershed 4), and Elkhead Creek and tributaries (Elkhead Creek subwatershed) (fig. 1). The Yampa River subwatershed 3 contains a large portion of the Yampa coal field that is in the UYRW (fig. 1).

For the UYRW and each subwatershed, the number of stream sites and number of samples with water-quality data for 1975 through 2009 are summarized in table 6. Counts are included for physical properties and chemical constituents

discussed in this report. The greatest number of sites sampled are in the Yampa River subwatershed 1, and the fewest are in the Elkhead Creek subwatershed. About 29 percent (62 of 211) of the sites are main-stem Yampa River sites. No sites were sampled in a large area in the northern one-third of the watershed or in some areas in the southern part of the watershed (fig. 3).

A total of 5,861 samples were collected from streams in the UYRW (table 6). Most samples have data for physical properties (97 percent) and streamflow (63 percent). Fewer data are available for dissolved solids (30 percent), major ions (33 percent), nutrients (37 percent), and trace elements (41 percent). The greatest number of samples was collected in Yampa River subwatershed 1 and the fewest in Yampa River subwatershed 4. Data were collected every year at only one site, Yampa River at Steamboat Springs (site 153), in Yampa River subwatershed 2 (appendix 3). The largest yearly gaps in data collection occurred in Yampa River subwatershed 4. For the UYRW and most subwatersheds, more samples were collected from May through July than in other months of the year. About 13 percent (27 of 211) of the sites had data for more than 50 samples; almost one-half (96 of 211) had data for 5 or fewer samples. Because of this large difference in sampling frequency at some sites, some results may be biased by data for sites with a high sample count (appendix 3). The CDPHE has an active sampling program, but the CDPHE data in STORET and in the UYRW water-quality database were only available through 2007.

Physical Properties

Physical properties analyzed for this study include specific conductance, pH, water temperature, dissolved oxygen, hardness, and acid neutralizing capacity (ANC). These data are summarized in table 8. A total of 5,660 samples from 209 stream sites have data for physical properties from 1975 through 2009 (table 6). The greatest number of samples analyzed for physical properties was collected during the 1990s. The greatest number of samples with data for physical properties were collected in Yampa River subwatershed 1, and the fewest were in Yampa River subwatershed 4.

Specific conductance is the ability of a substance to conduct an electric current (Hem, 1992). In water, it is proportional to the concentration of major dissolved constituents (bicarbonate, calcium, chloride, fluoride, magnesium, potassium, silica, sodium, and sulfate). The weathering of minerals in soil and bedrock is a primary source of major dissolved constituents to water. Atmospheric deposition can be a substantial source of dissolved chloride and sulfate particularly in areas with crystalline bedrock, which typically has low chlorine and sulfur content (Mast, 2007).

Specific conductance in the UYRW was measured in 4,305 samples from 1975 through 2009 and ranged from 2 to 10,000 microsiemens per centimeter at 25° Celsius (μS/cm) with a median of 315 μS/cm (table 8). Lower specific conductance values (less than 200 μS/cm) were most common in the

Table 6. Number of stream sites and samples with water-quality data for the Upper Yampa River watershed and subwatersheds, Colorado, by physical properties or constituent group, 1975 through 2009.

[No., number; --, no data]

Physical properties or constituent group	Upper Yampa River watershed		Subwatershed											
			Yampa River and tributaries upstream from Chuck Lewis State Wildlife Area (Yampa River subwatershed 1)		Yampa River and tributaries from Chuck Lewis State Wildlife Area to Elk River confluence (Yampa River subwatershed 2)		Elk River and tributaries (Elk River subwatershed)		Yampa River and tributaries from Elk River confluence to Town of Hayden (Yampa River subwatershed 3)		Yampa River and tributaries from Town of Hayden to Elkhead Creek confluence (Yampa River subwatershed 4)		Elkhead Creek and tributaries (Elkhead Creek subwatershed)	
	No. of sites	No. of samples	No. of sites	No. of samples	No. of sites	No. of samples	No. of sites	No. of samples	No. of sites	No. of samples	No. of sites	No. of samples	No. of sites	No. of samples
Physical properties	209	5,660	55	1,932	52	1,063	23	620	49	1,358	16	204	13	483
Streamflow (instantaneous)	139	3,684	35	801	35	795	15	467	35	1,060	11	137	8	424
Dissolved solids	110	1,743	28	429	13	166	17	200	33	647	10	101	9	200
Major ions	130	1,921	28	473	26	220	17	220	39	697	11	106	8	205
Nutrients	164	2,162	33	538	44	302	21	222	46	756	12	128	8	216
Trace elements	145	2,427	42	1,022	25	253	19	193	39	708	11	156	9	95
Uranium	14	51	4	21	1	4	4	12	4	13	1	1	--	--
Coliform bacteria	89	432	20	108	33	175	8	36	18	32	4	12	6	69
Suspended sediment	65	1,079	17	261	14	66	10	101	12	399	5	92	7	160
Total number of sites and samples	211	5,861	55	1,971	52	1,097	23	629	51	1,452	16	225	14	487

Table 7.   Period of water-quality record and number of water-quality samples collected per year from streams, by physical properties or constituent group, Upper Yampa River watershed and subwatersheds, Colorado, 1975 through 2009

[--, no data]

Physical properties or constituent group	Number of water-quality samples collected																
	1975	1976	1977	1978	1979	1980	1981	1982	1983	1984	1985	1986	1987	1988	1989	1990	1991
Upper Yampa River watershed																	
Physical properties	194	231	76	63	103	137	177	145	125	47	93	169	159	137	186	170	280
Streamflow, instantaneous	118	193	26	49	59	103	133	77	63	29	73	149	145	123	171	143	162
Dissolved solids	45	52	37	44	67	88	146	60	76	37	62	63	41	36	28	17	33
Major ions	50	56	37	46	67	89	148	61	77	38	63	64	41	36	32	15	34
Nutrients	137	156	36	46	64	109	146	79	76	38	63	64	41	35	46	27	37
Trace elements	82	101	36	47	77	109	88	95	59	37	51	52	33	32	22	22	79
Uranium	--	--	--	--	8	--	8	6	4	--	--	--	--	--	--	--	--
Coliform bacteria	56	69	--	--	--	--	--	--	--	--	--	--	--	--	18	8	19
Suspended sediment	42	137	23	45	57	98	69	65	37	13	29	69	42	39	34	19	32
Yampa River and tributaries upstream from Chuck Lewis State Wildlife Area (Yampa River subwatershed 1)																	
Physical properties	47	40	16	10	27	14	22	14	14	14	22	54	49	48	47	55	124
Streamflow, instantaneous	24	23	--	1	2	7	--	--	--	2	9	45	41	41	38	38	49
Dissolved solids	10	9	6	3	13	12	19	11	11	14	19	20	22	20	16	8	17
Major ions	10	9	6	4	10	12	20	11	11	14	19	21	22	20	16	7	17
Nutrients	21	27	6	4	10	12	20	11	12	14	19	21	22	19	30	18	25
Trace elements	20	17	6	4	14	11	13	11	11	13	14	16	15	16	16	17	44
Uranium	--	--	--	--	4	--	4	3	3	--	--	--	--	--	--	--	--
Coliform bacteria	15	15	--	--	--	--	--	--	--	--	--	--	--	--	18	8	14
Suspended sediment	17	21	--	6	13	--	--	--	--	--	--	42	39	39	28	16	25
Yampa River and tributaries from Chuck Lewis State Wildlife Area to Elk River confluence (Yampa River subwatershed 2)																	
Physical properties	47	75	9	6	5	7	10	31	42	10	20	55	49	26	59	31	56
Streamflow, instantaneous	29	46	--	--	--	--	5	6	19	10	19	52	49	26	59	32	46
Dissolved solids	2	--	--	--	--	--	5	3	15	5	10	8	1	--	1	--	5
Major ions	2	1	--	--	--	--	5	3	16	6	11	8	1	--	5	--	5
Nutrients	31	54	--	--	--	--	5	5	14	6	11	8	1	--	5	--	5
Trace elements	6	11	--	--	--	--	--	6	12	6	4	1	--	--	--	--	13
Uranium	--	--	--	--	--	--	--	--	--	--	--	--	--	--	--	--	--
Coliform bacteria	21	38	--	--	--	--	--	--	--	--	--	--	--	--	--	--	5
Suspended sediment	8	22	--	--	--	--	--	6	12	3	5	1	--	--	5	--	4
Elk River and tributaries (Elk River subwatershed)																	
Physical properties	16	23	10	2	7	11	13	12	9	1	2	6	8	15	29	36	30
Streamflow, instantaneous	11	17	1	--	--	--	--	--	4	1	2	6	8	15	29	36	31
Dissolved solids	4	6	--	--	4	6	6	3	4	--	--	--	--	--	--	3	3
Major ions	4	6	--	--	4	6	6	3	4	--	--	--	--	--	--	3	4
Nutrients	10	9	--	--	4	6	6	3	4	--	--	--	--	--	--	3	4
Trace elements	8	6	--	--	4	6	6	3	--	--	--	--	--	--	--	2	2
Uranium	--	--	--	--	2	--	2	1	--	--	--	--	--	--	--	--	--
Coliform bacteria	5	11	--	--	--	--	--	--	--	--	--	--	--	--	--	--	--
Suspended sediment	6	17	1	--	--	--	--	--	--	--	--	--	--	--	--	3	3
Yampa River and tributaries from Elk River confluence to Town of Hayden (Yampa River subwatershed 3)																	
Physical properties	68	72	36	35	48	68	90	65	52	21	47	46	43	41	40	35	44
Streamflow, instantaneous	43	91	25	32	47	66	83	57	37	15	41	38	37	34	34	28	26
Dissolved solids	25	34	31	32	39	46	81	42	42	18	33	35	18	16	11	6	8
Major ions	30	36	31	32	43	47	82	43	42	18	33	35	18	16	11	5	8
Nutrients	64	57	30	32	40	58	80	54	42	18	33	35	18	16	11	6	3
Trace elements	41	57	30	33	47	58	47	63	36	18	33	35	18	16	6	3	3
Uranium	--	--	--	--	2	--	2	2	1	--	--	--	--	--	--	--	--
Coliform bacteria	11	--	--	--	--	--	--	--	--	--	--	--	--	--	--	--	--
Suspended sediment	7	64	22	29	34	64	44	46	25	10	24	26	3	--	1	--	--

Table 7. Period of water-quality record and number of water-quality samples collected per year from streams, by physical properties or constituent group, Upper Yampa River watershed and subwatersheds, Colorado, 1975 through 2009.—Continued

[--, no data]

Physical properties or constituent group	Number of water-quality samples collected																	
	1992	1993	1994	1995	1996	1997	1998	1999	2000	2001	2002	2003	2004	2005	2006	2007	2008	2009
Upper Yampa River watershed																		
Physical properties	360	285	296	262	242	263	182	219	176	298	159	118	97	70	55	63	20	3
Streamflow, instantaneous	233	150	140	161	126	132	137	187	155	119	90	91	76	56	4	4	4	3
Dissolved solids	25	14	9	18	69	124	41	88	52	198	88	46	19	3	7	4	4	2
Major ions	25	14	9	18	69	124	41	89	61	199	89	49	31	8	55	63	20	3
Nutrients	31	11	13	19	69	126	41	89	61	199	89	42	23	8	55	63	20	3
Trace elements	122	106	124	108	127	139	65	61	47	184	87	56	31	8	55	63	20	2
Uranium	--	--	--	--	--	--	--	--	--	--	--	5	17	--	3	--	--	--
Coliform bacteria	18	4	4	2	7	10	9	22	14	63	45	10	3	4	4	20	20	3
Suspended sediment	19	--	--	8	22	28	24	54	23	14	12	25	--	--	--	--	--	--
Yampa River and tributaries upstream from Chuck Lewis State Wildlife Area (Yampa River subwatershed 1)																		
Physical properties	202	173	184	142	101	102	54	51	38	93	60	36	31	15	18	15	--	--
Streamflow, instantaneous	102	50	37	49	30	30	29	33	27	19	21	19	22	13	--	--	--	--
Dissolved solids	10	2	--	--	12	33	3	11	12	74	35	2	5	--	--	--	--	--
Major ions	10	2	--	--	12	33	3	11	12	74	35	4	15	--	18	15	--	--
Nutrients	20	2	--	--	12	34	3	11	12	74	35	4	7	--	18	15	--	--
Trace elements	99	93	109	87	61	70	22	22	16	74	43	20	15	--	18	15	--	--
Uranium	--	--	--	--	--	--	--	--	--	--	--	2	5	--	--	--	--	--
Coliform bacteria	14	--	--	--	--	--	--	8	--	7	9	--	--	--	--	--	--	--
Suspended sediment	15	--	--	--	--	--	--	--	--	--	--	--	--	--	--	--	--	--
Yampa River and tributaries from Chuck Lewis State Wildlife Area to Elk River confluence (Yampa River subwatershed 2)																		
Physical properties	60	32	49	42	42	38	38	31	26	31	25	17	19	20	8	24	20	3
Streamflow, instantaneous	57	32	49	42	18	22	27	29	24	17	17	16	16	16	4	4	4	3
Dissolved solids	4	3	--	--	6	21	5	9	7	18	12	5	5	3	3	4	4	2
Major ions	4	3	--	--	6	21	5	9	7	18	13	5	7	4	8	24	20	3
Nutrients	4	3	--	--	6	21	5	9	7	18	13	5	7	4	8	24	20	3
Trace elements	8	3	--	--	28	20	21	8	5	18	13	5	7	4	8	24	20	2
Uranium	--	--	--	--	--	--	--	--	--	--	--	1	3	--	--	--	--	--
Coliform bacteria	4	3	--	--	1	4	4	7	4	15	11	4	3	4	4	20	20	3
Suspended sediment	--	--	--	--	--	--	--	--	--	--	--	--	--	--	--	--	--	--
Elk River and tributaries (Elk River subwatershed)																		
Physical properties	28	23	14	11	17	31	20	63	44	46	29	38	12	1	9	4	--	--
Streamflow, instantaneous	28	23	14	11	11	15	19	62	44	19	20	30	9	1	--	--	--	--
Dissolved solids	5	4	4	2	11	18	1	36	13	27	11	22	3	--	4	--	--	--
Major ions	5	4	4	2	11	18	1	37	21	27	11	23	3	--	9	4	--	--
Nutrients	5	4	4	1	11	19	1	37	21	27	11	16	3	--	9	4	--	--
Trace elements	2	3	4	2	11	19	1	17	15	27	11	28	3	--	9	4	--	--
Uranium	--	--	--	--	--	--	--	--	--	--	--	1	3	--	3	--	--	--
Coliform bacteria	--	--	--	--	1	3	--	--	--	10	6	--	--	--	--	--	--	--
Suspended sediment	4	--	--	--	--	--	--	35	19	--	--	13	--	--	--	--	--	--
Yampa River and tributaries from Elk River confluence to Town of Hayden (Yampa River subwatershed 3)																		
Physical properties	49	43	35	36	38	59	46	45	32	55	5	1	5	24	18	16	--	--
Streamflow, instantaneous	36	36	28	29	24	31	38	36	29	22	--	--	--	17	--	--	--	--
Dissolved solids	6	5	5	6	13	28	8	11	4	33	5	1	5	--	--	--	--	--
Major ions	6	5	5	6	13	28	8	11	4	33	5	1	5	3	18	16	--	--
Nutrients	2	1	5	6	13	28	8	11	4	33	5	1	5	3	18	16	--	--
Trace elements	2	1	5	6	13	28	16	9	3	33	5	1	5	3	18	16	--	--
Uranium	--	--	--	--	--	--	--	--	--	--	--	1	5	--	--	--	--	--
Coliform bacteria	--	--	--	--	--	--	1	3	2	12	3	--	--	--	--	--	--	--
Suspended sediment	--	--	--	--	--	--	--	--	--	--	--	--	--	--	--	--	--	--

Table 7.   Period of water-quality record and number of water-quality samples collected per year from streams, by physical properties or constituent group, Upper Yampa River watershed and subwatersheds, Colorado, 1975 through 2009.—Continued

[--, no data]

Physical properties or constituent group	Number of water-quality samples collected																
	1975	1976	1977	1978	1979	1980	1981	1982	1983	1984	1985	1986	1987	1988	1989	1990	1991
Yampa River and tributaries from Town of Hayden to Elkhead Creek confluence (Yampa River subwatershed 4)																	
Physical properties	8	3	--	9	11	35	38	13	--	--	--	--	--	--	--	4	17
Streamflow, instantaneous	4	3	--	15	10	37	38	14	--	--	--	--	--	--	--	--	--
Dissolved solids	1	--	--	8	9	24	35	1	--	--	--	--	--	--	--	--	--
Major ions	1	--	--	9	10	24	35	1	--	--	--	--	--	--	--	--	--
Nutrients	6	3	--	9	10	33	35	6	--	--	--	--	--	--	--	--	--
Trace elements	2	3	--	9	10	34	22	12	--	--	--	--	--	--	--	--	17
Uranium	--	--	--	--	--	--	--	--	--	--	--	--	--	--	--	--	--
Coliform bacteria	2	--	--	--	--	--	--	--	--	--	--	--	--	--	--	--	--
Suspended sediment	--	--	--	10	10	34	25	13	--	--	--	--	--	--	--	--	--
Elkhead Creek and tributaries (Elkhead Creek subwatershed)																	
Physical properties	8	18	5	1	5	2	4	10	8	1	2	8	10	7	11	9	9
Streamflow, instantaneous	7	13	--	1	--	--	--	--	3	1	2	8	10	7	11	9	10
Dissolved solids	3	3	--	1	2	--	--	--	4	--	--	--	--	--	--	--	--
Major ions	3	4	--	1	--	--	--	--	4	--	--	--	--	--	--	--	--
Nutrients	5	6	--	1	--	--	--	--	4	--	--	--	--	--	--	--	--
Trace elements	5	7	--	1	2	--	--	--	--	--	--	--	--	--	--	--	--
Coliform bacteria	2	5	--	--	--	--	--	--	--	--	--	--	--	--	--	--	--
Suspended sediment	4	13	--	--	--	--	--	--	--	--	--	--	--	--	--	--	--

Elk River subwatershed and headwater tributaries in Yampa River subwatershed 2 (fig. 6A). The underlying geology of these two subwatersheds is composed mostly of igneous and metamorphic rocks (fig. 5) that are resistant to weathering. Less weathering results in fewer dissolved constituents and low specific conductance in stream water. Specific conductance values greater than 1,000 µS/cm were most common in Yampa River subwatersheds 3 and 4 (fig. 6A). Cretaceous-age sedimentary rocks underlying these subwatersheds (fig. 5) are susceptible to weathering, which results in an increase in dissolved constituents and specific conductance in stream water. Although the Elkhead Creek subwatershed is also underlain by sedimentary rocks, the median specific conductance (305 µS/cm) for this subwatershed is much lower than that for Yampa River subwatersheds 3 and 4 (802 and 1,008 µS/cm, respectively). Differences in sedimentary rocks underlying the subwatersheds could account for some of the difference in specific conductance values (fig. 5). Also, processes in Elkhead Creek Reservoir could reduce specific conductance at stream sites (sites 41 and 42) just downstream from the reservoir outlet from what would occur naturally. Median specific conductance for these sites was about 100 µS/cm lower than the median value for site 40, upstream from the reservoir. Throughout the UYRW, specific conductance typically was lowest during snowmelt runoff when dilute waters dominate streamflow and were greater during most other times of the year when groundwater dominates streamflow. This is illustrated by the monthly distribution of specific conductance values measured at Yampa River at Milner (site 151) (fig. 7A). This pattern is present for many dissolved constituents in water.

Specific conductance data for three sites met the statistical requirements for trends testing (see "Temporal Trend Analysis" subsection). A statistically significant (p-value 0.03) trend in a downward direction was identified for specific conductance at Yampa River at Steamboat Springs (site 153) for 1997 through 2008 (table 9). The rate of change in specific conductance was small, less than 3 percent per year (magnitude about 6 µS/cm per year). The median specific conductance at the site was 264 µS/cm.

The pH represents the effective concentration (activity) of hydrogen ions in water and is measured as the negative logarithm of the hydrogen ion activity (Hem, 1992). Pure water has a pH of 7. A value lower than 7 is considered acidic, and a value higher than 7 is considered alkaline or basic. Stream water not influenced by human activities typically has a pH that ranges from 6.5 to 8.5. The pH of a water-quality sample can be affected by biological activity in a stream, geology, precipitation, and human activities. Diurnal fluctuations in pH (lowest in the morning, highest in the late afternoon) result when algae take up dissolved carbon dioxide through photosynthesis during the day, which decreases the concentration of carbonic acid dissolved in water (increasing pH), and release carbon dioxide through respiration at night (decreasing pH) (Hem, 1992). Values of pH decrease in the presence of sulfide-bearing minerals (including pyrite) in rocks and soils as water and oxygen react with sulfur to form sulfuric acid. This can occur naturally in areas with mineralized bedrock or in areas with hard-rock mining. Hard-rock and coal mining can result in acidic drainage to a stream. The solubility and biological

Table 7. Period of water-quality record and number of water-quality samples collected per year from streams, by physical properties or constituent group, Upper Yampa River watershed and subwatersheds, Colorado, 1975 through 2009.—Continued

[--, no data]

Physical properties or constituent group	Number of water-quality samples collected																	
	1992	1993	1994	1995	1996	1997	1998	1999	2000	2001	2002	2003	2004	2005	2006	2007	2008	2009
Yampa River and tributaries from Town of Hayden to Elkhead Creek confluence (Yampa River subwatershed 4)																		
Physical properties	11	5	2	1	1	--	--	--	2	18	6	--	12	6	--	2	--	--
Streamflow, instantaneous	--	--	--	--	--	--	--	--	--	--	--	--	11	5	--	--	--	--
Dissolved solids	--	--	--	--	--	--	--	--	1	15	--	6	1	--	--	--	--	--
Major ions	--	--	--	--	--	--	--	--	1	15	6	--	1	1	--	2	--	--
Nutrients	--	--	--	--	--	--	--	--	1	15	6	--	1	--	1	2	--	--
Trace elements	11	5	2	1	1	--	--	--	1	16	6	--	1	1	--	2	--	--
Uranium	--	--	--	--	--	--	--	--	--	--	--	--	1	--	--	--	--	--
Coliform bacteria	--	--	--	--	--	--	--	--	--	5	5	--	--	--	--	--	--	--
Suspended sediment	--	--	--	--	--	--	--	--	--	--	--	--	--	--	--	--	--	--
Elkhead Creek and tributaries (Elkhead Creek subwatershed)																		
Physical properties	10	9	12	30	43	33	24	29	34	55	34	26	18	4	2	2	--	--
Streamflow, instantaneous	10	9	12	30	43	34	24	27	31	42	32	26	18	4	--	--	--	--
Dissolved solids	--	--	--	10	27	24	24	21	15	31	19	16	--	--	--	--	--	--
Major ions	--	--	--	10	27	24	24	21	16	32	19	16	--	--	2	2	--	--
Nutrients	--	1	4	12	27	24	24	21	16	32	19	16	--	--	2	2	--	--
Trace elements	--	1	4	12	13	2	5	5	7	16	9	2	--	--	2	2	--	--
Coliform bacteria	--	1	4	2	5	3	4	4	8	14	11	6	--	--	--	--	--	--
Suspended sediment	--	--	--	8	22	28	24	19	4	14	12	12	--	--	--	--	--	--

availability of nutrients and trace elements and some chemical processes are pH dependent. A pH less than 4 and greater than 10 can affect the survivability of aquatic organisms (*http://extension.usu.edu/files/publications/publication/nr_wq_2005-19.pdf*, accessed June 2012).

For the UYRW, pH was measured in 3,140 samples from 1975 through 2009 at 184 sites. Values ranged from 2.1 to 9.8, and the median was 8.1 (table 8). Median pH was highest in the Yampa River subwatershed 1 and Elkhead Creek subwatershed (8.3 and 8.2, respectively) and lowest (7.6) in the Elk River subwatershed (fig. 6B). About 84 percent of pH values were between 6.5 and 8.5. Individual values greater than 8.5 were more common than individual values less than 6.5. Many high values occurred during summer afternoons. Values greater than 8.5 also were more common in samples from Yampa River subwatershed 3 than from the other subwatersheds. Fewer than 3 percent of sites had pH values that were not in attainment of the CDPHE aquatic-life standard for pH (fig. 8, table 10). These data were collected at four sites in Yampa River subwatershed 2 and one site in the Elk River subwatershed. Data for the four sites with pH not in attainment of the 6.5 standard were collected before 1988; current (2010) data were not available to evaluate pH at these sites. Nonattainment of the 9.0 standard for Yampa River above Elk River (site 145) was based on the three values collected at the site; collections dates were from 1999 through 2002. For pH and other water-quality constituents, the Yampa River at Steamboat Springs (site 153) was the only site with data that met the requirements for temporal trend analysis. No statistically significant (p-value 0.41) trend in pH was identified at the site for the period 1997 through 2008 (table 9).

Water temperature is an important property that controls biological and chemical reaction rates. Temperature often directly affects dissolved oxygen levels and life cycles of aquatic organisms (Hem, 1992). Water-temperature data in the UYRW database are instantaneous measurements made when a water-quality sample was collected. Water temperature was measured for 5,187 samples from 1975 through 2009 and ranged from –3 to 30.4 °C with a median of 7.5 °C (table 8). The Elk River subwatershed had the lowest median water temperature and the Yampa River subwatershed 4 had the highest, which are likely due to differences in the altitude and climate of sites where water temperatures were measured. Most (41 of 55) temperature values less than 0 °C were measured from November through February. All values greater than 25 °C were measured from June through August. Median values of water temperature for main-stem Yampa River sites were highest for Yampa River subwatersheds 3 and 4 and lowest for Yampa River subwatersheds 1 and 2 because of the inflow of streams in the latter two subwatersheds that drain high altitude, colder areas of the UYRW. The Yampa River at Steamboat Springs (site 153) was the only site with sufficient instantaneous water temperature data for temporal trend analysis; a statistically significant trend was not identified for 1997 through 2008 (table 9).

The Yampa River at Steamboat Springs site (site 153) also was the only site with a continuous record of water temperature; measurements were made at 15-minute intervals for most days from July 26, 2002, through April 13, 2005. Values greater than the June through September acute DM and chronic MWAT cold-water aquatic-life standards at this site occurred most often from mid-July through August during 2002, 2003, and

Table 8.   Summary statistics for selected physical property water-quality data and Colorado Department of Public Health and Environment in-stream water-quality standards for stream sampling sites in the Upper Yampa River watershed and subwatersheds, Colorado, 1975 through 2009.

[No., number; µS/cm; microsiemens per centimeter at 25 degrees Celsius; --, no water-quality standard; C, degrees Celsius; mg/L, milligrams per liter; CaCO$_3$, calcium carbonate. Number of significant figures for individual constituents may vary because data are from different sources and analytical periods. Water-quality standards are from Colorado Department of Public Health and Environment (2009a, 2010). See table 10 for sites in the Upper Yampa River watershed and subwatersheds with data for physical properties exceeding in-stream water-quality standards. Descriptions of stream segments are in Colorado Department of Public Health and Environment (2010)]

Physical property (reporting units)	No. of sites	No. of samples	No. of censored values[1]	Minimum value[2]	Median value	Maximum value	In-stream water-quality standard	No. of sites with data not in attainment of standard
Upper Yampa River watershed								
Specific conductance (µS/cm)	191	4,305	33	2	315	10,000	--	--
pH (standard units)	184	3,140	0	2.1	8.1	9.8	6.5–9.0	5
Water temperature (°C)	191	5,187	0	–3	7.5	30.4	([3])	[4]1
Oxygen, dissolved (mg/L)	171	2,797	0	1.2	9.8	17.8	[5]5.0, 6.0 (minima)	2
Hardness (mg/L as CaCO$_3$)	142	2,434	2	4	162	4,000	--	--
Acid neutralizing capacity (mg/L as CaCO$_3$)	134	2,244	1	2	136	660	--	--
Yampa River and tributaries upstream from Chuck Lewis State Wildlife Area (Yampa River subwatershed 1; stream segments 2a, 2c, 3, 4, 5, 6, 7)								
Specific conductance (µS/cm)	46	958	0	29	425	1,179	--	--
pH (standard units)	51	1,171	0	6.3	8.3	9.4	6.5–9.0	0
Water temperature (°C)	50	1,582	0	–3	8.0	28.0	([3])	--[4]
Oxygen, dissolved (mg/L)	51	1,293	0	1.2	9.8	15.0	6.0 (minima)	2
Hardness (mg/L as CaCO$_3$)	41	1,018	0	25	190	1,550	--	--
Acid neutralizing capacity (mg/L as CaCO$_3$)	41	989	0	4	152	420	--	--
Yampa River and tributaries from Chuck Lewis State Wildlife Area to Elk River confluence (Yampa River subwatershed 2; stream segments 2c, 3, 20a)								
Specific conductance (µS/cm)	48	968	30	2	57	750	--	--
pH (standard units)	39	288	0	5.6	7.8	9.7	6.5–9.0	4
Water temperature (°C)	42	1,003	0	–0.2	6.0	26.6	([3])	[4]1
Oxygen, dissolved (mg/L)	38	299	0	4.0	10.0	17.8	6.0 (minima)	0
Hardness (mg/L as CaCO$_3$)	26	218	1	4.3	48	320	--	--
Acid neutralizing capacity (mg/L as CaCO$_3$)	21	187	0	2	60	240	--	--
Elk River and tributaries (Elk River subwatershed; stream segments 8, 20a)								
Specific conductance (µS/cm)	22	570	3	16	80	750	--	--
pH (standard units)	22	233	0	6.0	7.6	9.4	6.5–9.0	1
Water temperature (°C)	23	613	0	–0.3	5.7	25.5	([3])	--[4]
Oxygen, dissolved (mg/L)	22	200	0	6.9	9.9	13.8	6.0 (minima)	0
Hardness (mg/L as CaCO$_3$)	17	209	1	4	21	120	--	--
Acid neutralizing capacity (mg/L as CaCO$_3$)	15	133	1	4	30	86	--	--
Yampa River and tributaries from Elk River confluence to Town of Hayden (Yampa River subwatershed 3; stream segments 2c, 11, 12, 13a, 13b, 13c, 13e, 13f, 20a)								
Specific conductance (µS/cm)	48	1,196	0	49	802	6,360	--	--
pH (standard units)	48	1,105	0	2.1	8.1	9.8	6.5–9.0	0
Water temperature (°C)	48	1,314	0	–0.2	8.9	28.6	([3])	--[4]
Oxygen, dissolved (mg/L)	38	668	0	3.0	9.7	16.0	[5]5.0, 6.0 (minima)	0
Hardness (mg/L as CaCO$_3$)	37	679	0	20	260	3,000	--	--
Acid neutralizing capacity (mg/L as CaCO$_3$)	37	667	0	8.2	162	660	--	--

Table 8. Summary statistics for selected physical property water-quality data and Colorado Department of Public Health and Environment in-stream water-quality standards for stream sampling sites in the Upper Yampa River watershed and subwatersheds, Colorado, 1975 through 2009.—Continued

[No., number; µS/cm; microsiemens per centimeter at 25 degrees Celsius; --, no water-quality standard; C, degrees Celsius; mg/L, milligrams per liter; $CaCO_3$, calcium carbonate. Number of significant figures for individual constituents may vary because data are from different sources and analytical periods. Water-quality standards are from Colorado Department of Public Health and Environment (2009a, 2010). See table 10 for sites in the Upper Yampa River watershed and subwatersheds with data for physical properties exceeding in-stream water-quality standards. Descriptions of stream segments are in Colorado Department of Public Health and Environment (2010)]

Physical property (reporting units)	No. of sites	No. of samples	No. of censored values[1]	Minimum value[2]	Median value	Maximum value	In-stream water-quality standard	No. of sites with data not in attainment of standard
Yampa River and tributaries from Town of Hayden to Elkhead Creek confluence (Yampa River subwatershed 4; stream segments 2c, 12, 13d)								
Specific conductance (µS/cm)	15	150	0	72.2	1,008	10,000	--	--
pH (standard units)	13	122	0	6.5	8.0	8.8	6.5–9.0	0
Water temperature (°C)	16	200	0	–0.3	9.5	25.2	([3])	--[4]
Oxygen, dissolved (mg/L)	11	123	0	4.5	9.2	14.0	[5]5.0, 6.0 (minima)	0
Hardness (mg/L as $CaCO_3$)	11	104	0	23	540	4,000	--	--
Acid neutralizing capacity (mg/L as $CaCO_3$)	12	145	0	12	230	590	--	--
Elkhead Creek and tributaries (Elkhead Creek subwatershed; stream segments 14, 15, 20b)								
Specific conductance (µS/cm)	12	463	0	100	305	912	--	--
pH (standard units)	11	221	0	7.3	8.2	9.5	6.5–9.0	0
Water temperature (°C)	12	475	0	–0.1	8.7	30.4	([3])	--[4]
Oxygen, dissolved (mg/L)	11	214	0	4.7	10.1	14.5	[5]5.0, 6.0 (minima)	0
Hardness (mg/L as $CaCO_3$)	10	206	0	47.4	128	331	--	--
Acid neutralizing capacity (mg/L as $CaCO_3$)	8	123	0	48.6	114	219	--	--

[1]Censored values can be expressed as values less than the laboratory reporting level.

[2]For some constituents with censored data, the minimum censored value is greater than the minimum detected value that is shown.

[3]See appendix 2 for Colorado Department of Public Health and Environment temperature standards for cold- and warm-water streams.

[4]Only one site, Yampa River at Steamboat Springs (site 153) in stream segment 2c, had data for the continuous monitoring of temperature. The temperature standards only were applied to temperature data for this site. See figure 9 and table 11 for exceedances of the standards.

[5]Water-quality standard varies by stream segment. See Colorado Department of Public Health and Environment (2010).

2004; and a few exceedances occurred during early September (figs. 9A, 9B). The June through September chronic MWAT was exceeded on the greatest number of sample days (table 11). Although some water temperature values were greater than the June through September standards, this does not necessarily mean that the values were not in attainment of the standards for regulatory purposes because streamflow was lower than normal during the measuring period. Water temperature may exceed the standards when streamflow is less than critical low flows calculated from CDPHE formulas (Colorado Department of Public Health and Environment, 2009a). Critical low flows were not calculated for this study. Water temperatures also did not meet the October through May acute DM and chronic MWAT cold-water standards, primarily during October (figs. 9C, 9D; table 11). In total, the DM and MWAT standards were exceeded on 16 and 26 percent of all sample days, respectively, possibly due to reduced streamflow during drought conditions and from upstream hydrologic modifications, streams depths that are less than those normally observed during summer, and (or)

changes in the river channel. The stream segment Yampa River from Oak Creek to Elkhead Creek, including Yampa River at Steamboat Springs (site 153), is on the CDPHE 2012 monitoring and evaluation list for water temperature (table 5) (Colorado Department of Public Health and Environment, 2012). Increases in water temperature on the main stem of the Yampa River during summer in combination with low streamflow and low dissolved oxygen concentrations can result in reduced habitat availability for fish species and can create less than optimal resource conditions.

Dissolved oxygen is the measurement of the oxygen in water that is available to fish and aquatic life. It varies with temperature, altitude, and water depth and is affected by many factors, such as photosynthesis, respiration activity, and inputs from point and nonpoint sources. Dissolved oxygen often has seasonal as well as diurnal variations. Concentrations are typically higher when water temperatures are colder (Hem, 1992).

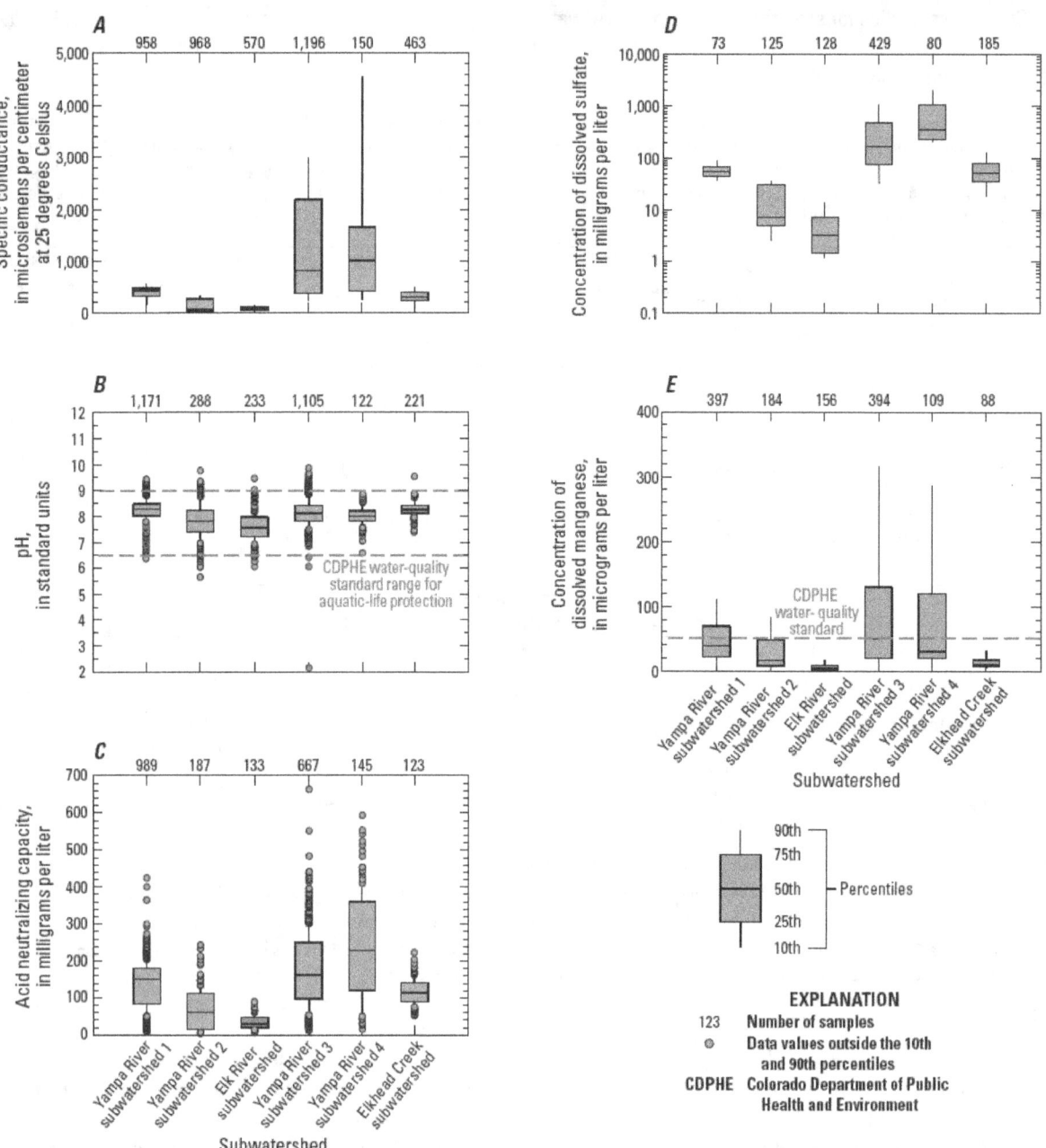

Figure 6. Distribution of (*A*) specific conductance, (*B*), pH, (*C*) acid neutralizing capacity, and concentrations of (*D*) dissolved sulfate and (*E*) dissolved manganese in stream water, by subwatershed, Upper Yampa River watershed, Colorado, 1975 through 2009.

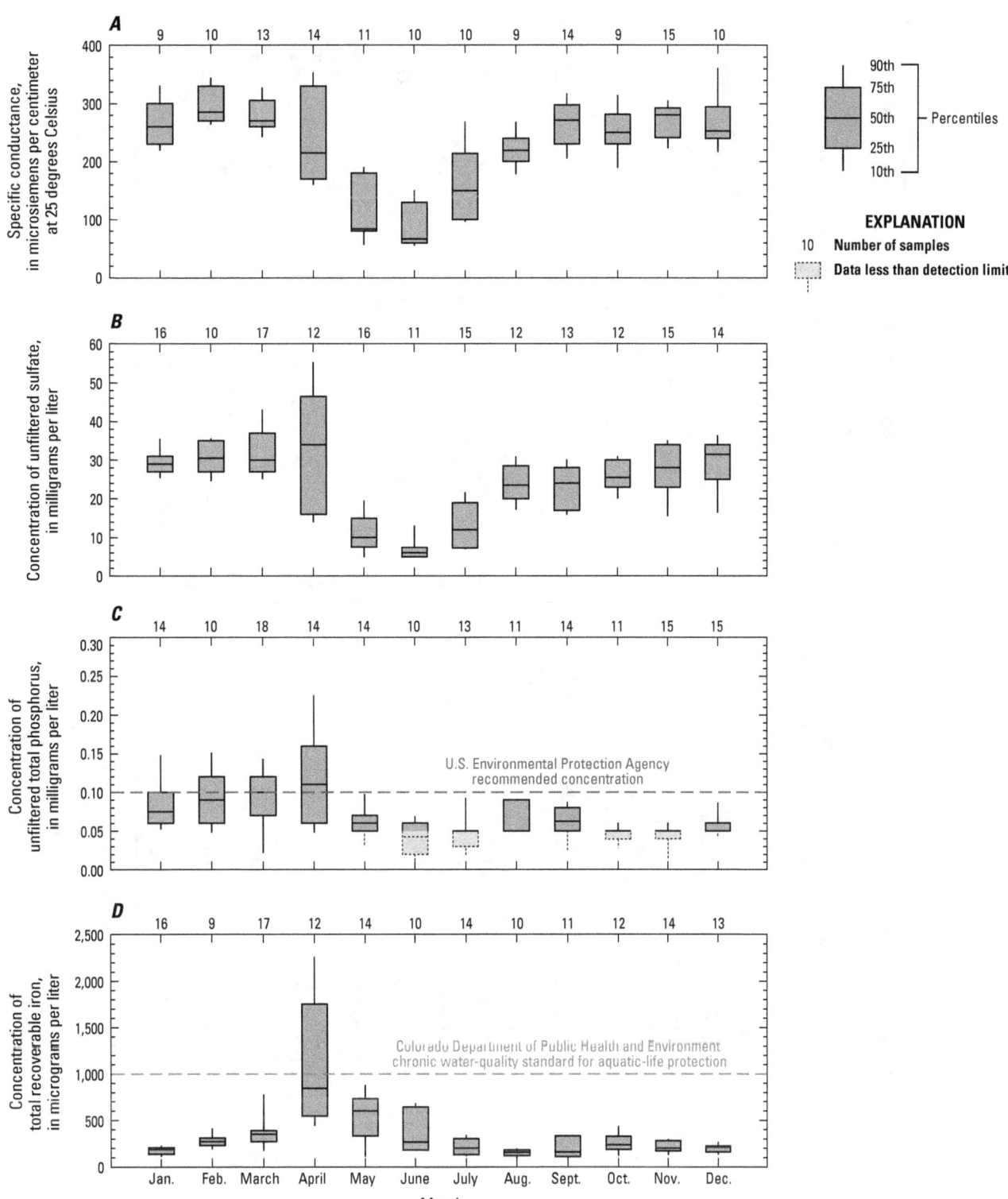

Figure 7. Distribution of (*A*) specific conductance, and concentrations of (*B*) unfiltered sulfate, (*C*) unfiltered total phosphorus, and (*D*) total recoverable iron at Yampa River at Milner (site 151), by month, Upper Yampa River watershed, Colorado, 1975 through 2007.

Table 9. Results of Seasonal Kendall trend analysis for selected physical properties or water-quality constituents at three stream sites in the Upper Yampa River watershed, Colorado, 1997 through 2008.

[No., number; p-value, significance of trend; μS/cm, microsiemens per centimeter at 25 degrees Celsius; --, trend not statistically significant; C, degrees Celsius; mg/L, milligrams per liter; CaCO$_3$, calcium carbonate; P, phosphorus; μg/L, micrograms per liter; col/100 mL, colonies per 100 milliliters. Results that are statistically significant are shown in **bold type**. See figure 3 and appendix 3 for location of sites and additional site information]

Physical property or constituent (reporting units)	Period of water-quality record (calendar year)	No. of samples/ No. of censored values	Flow adjustment	p-value	Trend direction	Trend slope	Magnitude of trend slope (percent/year)	Magnitude of trend slope (units/year)	Median value
Yampa River above Stagecoach Reservoir									
(site 147 in Yampa River subwatershed 1, Yampa River and tributaries upstream from Chuck Lewis State Wildlife Area)									
Specific conductance (μS/cm)	1990–2003	167/0	Yes	0.77	None	--	--	--	452
Elk River near Milner, CO									
(site 33 in Elk River subwatershed, Elk River and tributaries)									
Specific conductance (μS/cm)	1990–2003	162/0	Yes	0.08	None	--	--	--	122
Yampa River at Steamboat Springs									
(site 153 in Yampa River subwatershed 2, Yampa River and tributaries from Chuck Lewis State Wildlife Area to Elk River confluence)									
Specific conductance (μS/cm)	1997–2008	127/0	Yes	**0.03**	**Down**	**-0.028**	**-2.7**	**-6.0**	264
pH (standard units)	1997–2008	55/0	Yes	0.41	--	--	--	--	8.2
Water temperature (°C)	1997–2008	125/0	Yes	0.17	--	--	--	--	6.0
Hardness (mg/L as CaCO$_3$)	1997–2008	49/0	Yes	0.09	--	--	--	--	116
Dissolved solids (mg/L)	1997–2008	44/0	Yes	**0.03**	**Down**	**-0.030**	**-3.0**	**-3.9**	153
Calcium, dissolved (mg/L)	1997–2008	49/0	Yes	0.07	--	--	--	--	30.0
Magnesium, dissolved (mg/L)	1997–2008	49/0	Yes	0.08	--	--	--	--	9.84
Sodium, dissolved (mg/L)	1997–2008	49/0	Yes	0.40	--	--	--	--	8.20
Potassium, dissolved (mg/L)	1997–2008	49/0	Yes	0.26	--	--	--	--	1.74
Sulfate, dissolved (mg/L)	1997–2008	48/0	Yes	0.06	--	--	--	--	27.4
Chloride, dissolved (mg/L)	1997–2008	48/0	Yes	0.06	--	--	--	--	3.64
Silica, dissolved (mg/L)	1997–2008	48/0	Yes	0.77	--	--	--	--	9.72
Total phosphorous, unfiltered (mg/L as P)	1997–2008	48/3	Yes	**0.04**	**Up**	**0.031**	**3.1**	**0.001**	0.036
Iron, total recoverable (μg/L)	1997–2008	48/0	Yes	0.53	--	--	--	--	255
Manganese, dissolved (μg/L)	1997–2008	48/0	Yes	0.17	--	--	--	--	17.4
Manganese, total recoverable (μg/L)	1997–2008	48/0	Yes	**0.03**	**Down**	**-0.047**	**-4.6**	**-2.6**	45.6
Escherichia coli (col/100 mL)	1997–2008	47/4	No	0.06	--	--	--	--	2.7

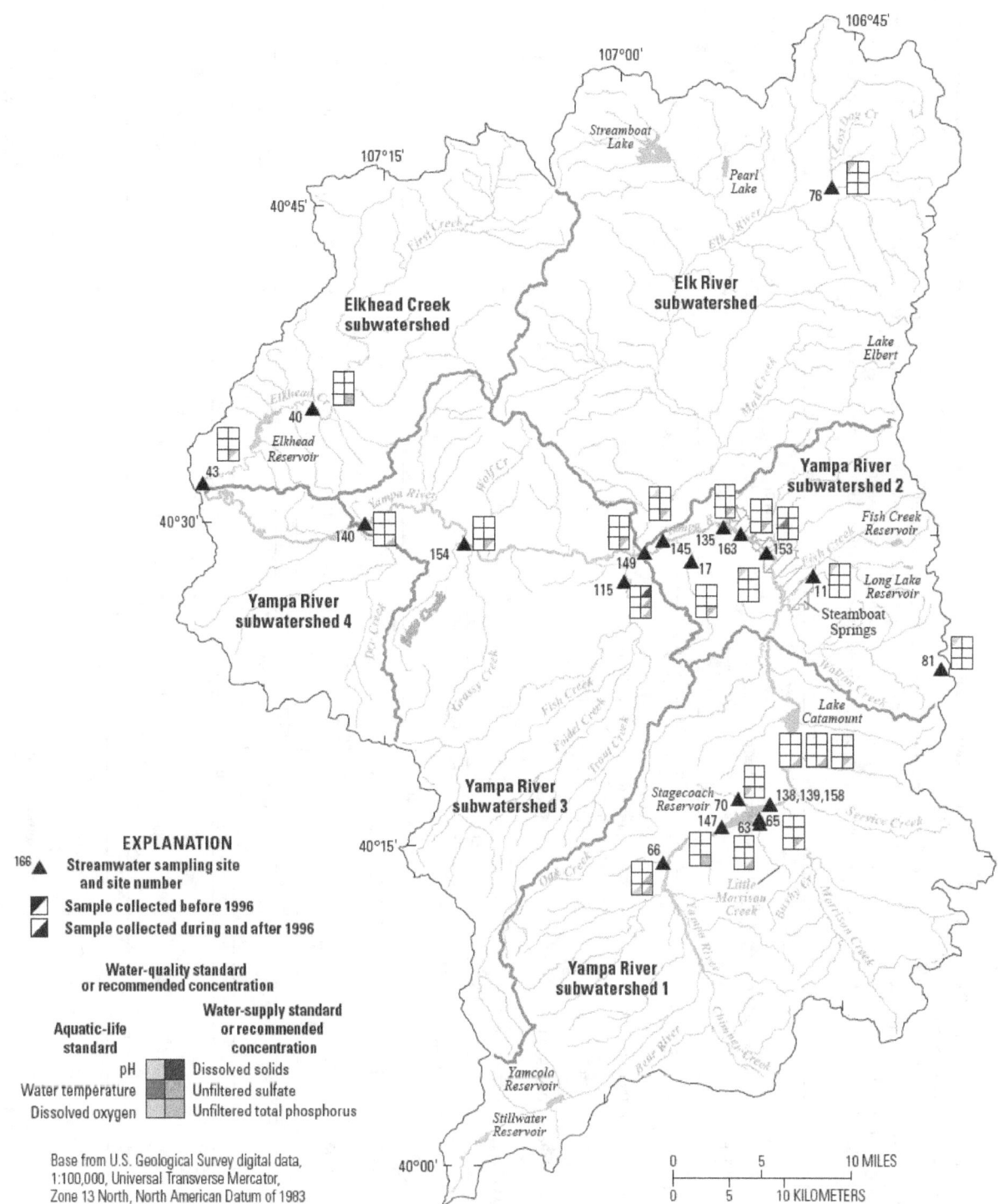

Figure 8.   Location of water-quality sampling sites with exceedances of Colorado Department of Public Health and Environment in-stream water-quality standards for physical properties (pH, water temperature, dissolved oxygen), dissolved solids, and unfiltered sulfate and U.S. Environmental Protection Agency recommended concentrations for total phosphorus, Upper Yampa River watershed, Colorado, 1975 through 2009.

Table 10. Stream water-quality sites with data in exceedance of Colorado Department of Public Health and Environment in-stream water-quality standards for selected physical properties and water-quality constituents and U.S. Environmental Protection Agency recommended concentrations for total phosphorus, Upper Yampa River watershed, Colorado, 1975 through 2009.

[No, number; Min., minimum; Max., maximum; WS, water supply; SMCL, secondary maximum contaminant level; USEPA, U.S. Environmental Protection Agency; <, less than. Subwatershed definitions: Yampa River subwatershed 1, Yampa River and tributaries upstream from Chuck Lewis State Wildlife Area; Yampa River subwatershed 2, Yampa River and tributaries from Chuck Lewis State Wildlife Area to Elk River confluence; Elk River subwatershed, Elk River and tributaries; Yampa River subwatershed 3, Yampa River and tributaries from Elk River confluence to Town of Hayden; Yampa River subwatershed 4, Yampa River and tributaries from Town of Hayden to Elkhead Creek confluence; Elkhead Creek subwatershed, Elkhead Creek and tributaries. Water-quality standards are from Colorado Department of Public Health and Environment (2009a, 2010); standards are for protection of aquatic life, unless otherwise stated. Recommended concentrations for total phosphorus are from U.S. Environmental Protection Agency (2000). Nonattainment of a standard is based on the 15th, 50th, or 85th percentile value shown in the table. See appendix 3 for additional site information. Descriptions of stream segments are in Colorado Department of Public Health and Environment (2010)]

Site no. (see figures 8, 12)	Site name in Upper Yampa River watershed water-quality database	Site identifier	Stream segment	Subwatershed	Period of water-quality record (calendar year)	No. of samples	No. of censored values[1]	Min. value	Max. value	Percentile value (percentile)	In-stream water-quality standard or recommended value
					pH (standard units)						
11	BURGESS CREEK AB SKI AREA NR STEAMBOAT SPGS, CO.	402802106471000	3	Yampa River subwatershed 2	1975–76	6	0	6.2	7.4	6.4 (15th)	6.5
53	FISH CREEK NEAR STEAMBOAT SPRINGS, CO	09239000	3	Yampa River subwatershed 2	1975–76	6	0	6.0	6.8	6.0 (15th)	6.5
76	NORTH FORK ELK RIVER NEAR HINMAN PARK, CO.	404620106462200	20a	Elk River subwatershed	1975–76	3	0	6.2	7.1	6.4 (15th)	6.5
81	NORTH FORK WALTON CREEK NR RABBIT EARS PASS, CO.	09238300	3	Yampa River subwatershed 2	[2]1975–87	10	0	5.6	8.3	6.4 (15th)	6.5
145	YAMPA RIVER ABV ELK RIVER	402936106565000	2c	Yampa River subwatershed 2	[2]1999–2002	3	0	9.2	9.2	9.2 (85th)	9.0
					Water temperature (degrees Celsius)						
153	YAMPA RIVER AT STEAMBOAT SPRINGS, CO	09239500	2c	Yampa River subwatershed 2	[3]2002–05	[3]1,802	0	0.0	31.3	—[3]	—[3]
					Dissolved oxygen (milligrams per liter)						
66	LITTLE WHITE SNAKE CK @ HWY 131	12897	4	Yampa River subwatershed 1	2001, 2006–07	4	0	2.2	6.2	3.8 (15th)	6.0 (minima)
70	MARTIN C AB DAM SITE NR OAK CREEK, CO	401729106514601	3	Yampa River subwatershed 1	1986–88	14	0	5.0	11.6	5.7 (15th)	6.0 (minima)
					Dissolved solids (milligrams per liter)[4]						
115	TROUT CK NR. MOUTH	12876	13f	Yampa River subwatershed 3	[2]1996–2004	17	0	160	680	636 (85th)	500 WS (SMCL)
					Sulfate, unfiltered (milligrams per liter)						
66	LITTLE WHITE SNAKE CK @ HWY 131	12897	4	Yampa River subwatershed 1	2001, 2006–07	4	0	210	290	285 (85th)	250 WS (SMCL)
115	TROUT CK NR. MOUTH	12876	13f	Yampa River subwatershed 3	[2]1996–2007	22	0	49	310	298 (85th)	250 WS (SMCL)
					Total phosphorus, unfiltered (milligrams per liter)[5]						
17	COW CR. NR STEAMBOAT SPRINGS,CO.	402836106550100	3	Yampa River subwatershed 2	1982	3	0	0.053	0.921	0.674 (85th)	0.1 (USEPA recommended)
40	ELKHEAD CREEK ABOVE LONG GULCH, NEAR HAYDEN, CO	09246200	15	Elkhead Creek subwatershed	1995–2003	83	20	<0.010	0.923	0.132 (85th)	0.05 (USEPA recommended)
43	ELKHEAD CREEK NEAR CRAIG, CO	09246500	15	Elkhead Creek subwatershed	[2]1975–78	6	1	<0.01	0.17	0.11 (85th)	0.1 (USEPA recommended)

Table 10. Stream water-quality sites with data in exceedance of Colorado Department of Public Health and Environment in-stream water-quality standards for selected physical properties and water-quality constituents and U.S. Environmental Protection Agency recommended concentrations for total phosphorus, Upper Yampa River watershed, Colorado, 1975 through 2009.—Continued

[No., number; Min., minimum; Max., maximum; WS, water supply; SMCL, secondary maximum contaminant level; USEPA, U.S. Environmental Protection Agency; <, less than. Subwatershed definitions: Yampa River subwatershed 1, Yampa River and tributaries upstream from Chuck Lewis State Wildlife Area; Yampa River subwatershed 2, Yampa River and tributaries from Chuck Lewis State Wildlife Area to Elk River confluence; Elk River subwatershed, Elk River and tributaries; Yampa River subwatershed 3, Yampa River and tributaries from Elk River confluence to Town of Hayden; Yampa River subwatershed 4, Yampa River and tributaries from Town of Hayden to Elkhead Creek confluence; Elkhead Creek subwatershed, Elkhead Creek and tributaries. Water-quality standards are from Colorado Department of Public Health and Environment (2009a, 2010); standards are for protection of aquatic life, unless otherwise stated. Recommended concentrations for total phosphorus are from U.S. Environmental Protection Agency (2000). Nonattainment of a standard is based on the 15th, 50th, or 85th percentile value shown in the table. See appendix 3 for additional site information. Descriptions of stream segments are in Colorado Department of Public Health and Environment (2010)]

Site no. (see figures 8, 12)	Site name in Upper Yampa River watershed water-quality database	Site identifier	Stream segment	Subwatershed	Period of water-quality record (calendar year)	No. of samples	No. of censored values[1]	Min. value	Max. value	Percentile value (percentile)	In-stream water-quality standard or recommended value
					Total phosphorus, unfiltered (milligrams per liter)[5]—Continued						
63	LITTLE MORRISON CK @ RD 18A	12896	3	Yampa River subwatershed 1	[2]2001–07	7	0	0.06	0.13	0.12 (85th)	0.05 (USEPA recommended)
65	LITTLE MORRISON CREEK NEAR STAGECOACH, CO.	401634106502200	3	Yampa River subwatershed 1	1975–76	4	0	0.12	0.51	0.36 (85th)	0.05 (USEPA recommended)
66	LITTLE WHITE SNAKE CK @ HWY 131	12897	4	Yampa River subwatershed 1	2001, 2006–07	4	0	0.09	0.56	0.40 (85th)	0.1 (USEPA recommended)
115	TROUT CK NR. MOUTH	12876	13f	Yampa River subwatershed 3	[2]1996–07	22	0	0.02	0.40	0.13 (85th)	0.1 (USEPA recommended)
135	YAMPA R. BL KOA CAMPGROUNDS NR STEAMBOAT SPGS, CO.	403017106525800	2c	Yampa River subwatershed 1	1975–76	4	0	0.11	0.18	0.16 (85th)	0.1 (USEPA recommended)
138	YAMPA R. BLW STAGECOACH RES.	12808	2a	Yampa River subwatershed 2	1996–99	14	0	0.03	0.13	0.12 (85th)	0.1 (USEPA recommended)
139	YAMPA R. D/S STAGECOACH RES. DAM	12808P	2a	Yampa River subwatershed 1	[2]2000–07	21	0	0.04	0.15	0.11 (85th)	0.1 (USEPA recommended)
140	YAMPA R. N. OF HAYDEN @ CALIFORNIA PARK RD	12802	2c	Yampa River subwatershed 3	[2]1996–99	13	0	0.03	0.26	0.12 (85th)	0.1 (USEPA recommended)
145	YAMPA RIVER ABOVE ELK RIVER NEAR MILNER, CO.	402932106564900	2c	Yampa River subwatershed 2	1975–76	9	0	0.03	0.17	0.16 (85th)	0.1 (USEPA recommended)
147	YAMPA RIVER ABOVE STAGECOACH RESERVOIR, CO	09237450	2a	Yampa River subwatershed 1	1988–92	24	0	0.02	0.20	0.11 (85th)	0.05 (USEPA recommended)
147	YAMPA R. U/S STAGECOACH RES @ CR16	12809	2a	Yampa River subwatershed 1	[2]1999–07	22	1	<0.01	0.10	0.08 (85th)	0.05 (USEPA recommended)
149	YAMPA RIVER AT ELK RIVER JUNCTION NR MILNER, CO.	402902106580000	2c	Yampa River subwatershed 2	1975–76	4	0	0.09	0.15	0.15 (85th)	0.1 (USEPA recommended)
154	YAMPA RIVER BELOW DIVERSION, NEAR HAYDEN, CO.	09244410	2c	Yampa River subwatershed 3	1975–82	81	0	0.01	0.24	0.12 (85th)	0.1 (USEPA recommended)
158	YAMPA RIVER BELOW STAGECOACH RESERVOIR, CO	09237500	2a	Yampa River subwatershed 1	1988–92	24	0	0.03	0.30	0.18 (85th)	0.1 (USEPA recommended)
163	YAMPA RIVER BELOW STEAMBOAT II SEWAGE PLANT, CO.	403002106545500	2c	Yampa River subwatershed 2	1975–76	4	0	0.09	0.24	0.20 (85th)	0.1 (USEPA recommended)

Table 10. Stream water-quality sites with data in exceedance of Colorado Department of Public Health and Environment in-stream water-quality standards for selected physical properties and water-quality constituents and U.S. Environmental Protection Agency recommended concentrations for total phosphorus, Upper Yampa River watershed, Colorado, 1975 through 2009.—Continued

[No., number; Min., minimum; Max., maximum; WS, water supply; SMCL, secondary maximum contaminant level; USEPA, U.S. Environmental Protection Agency; <, less than. Subwatershed definitions: Yampa River subwatershed 1, Yampa River and tributaries upstream from Chuck Lewis State Wildlife Area; Yampa River subwatershed 2, Yampa River and tributaries from Chuck Lewis State Wildlife Area to Elk River confluence; Elk River subwatershed, Elk River and tributaries; Yampa River subwatershed 3, Yampa River and tributaries from Elk River confluence to Town of Hayden; Yampa River subwatershed 4, Yampa River and tributaries from Town of Hayden to Elkhead Creek confluence; Elkhead Creek subwatershed, Elkhead Creek and tributaries. Water-quality standards are from Colorado Department of Public Health and Environment (2009a, 2010); standards are for protection of aquatic life, unless otherwise stated. Recommended concentrations for total phosphorus are from U.S. Environmental Protection Agency (2000). Nonattainment of a standard is based on the 15th, 50th, or 85th percentile value shown in the table. See appendix 3 for additional site information. Descriptions of stream segments are in Colorado Department of Public Health and Environment (2010)]

Site no. (see figures 8, 12)	Site name in Upper Yampa River watershed water-quality database	Site identifier	Stream segment	Subwatershed	Period of water-quality record (calendar year)	No. of samples	No. of censored values[1]	Min. value	Max. value	Percentile value (percentile)	In-stream water-quality standard or recommended value
	Copper, dissolved (micrograms per liter)										
45	ENGLISH CREEK ABOVE MOUTH, NEAR CLARK, CO.	404727106453700	20a	Elk River subwatershed	[2]1999–2003	7	4	[6]0.566	3.21	2.24 (85th)	2.02 (acute)/1.61 (chronic)
49	FISH CR AT UPPER STA NR STEAMBOAT SPRINGS, CO	09238900	3	Yampa River subwatershed 2	1984–85	3	1	<1	4	4 (85th)	1.72 (acute)/1.39 (chronic)
101	SAGE CREEK NEAR HAYDEN, CO.	402918107094400	13e	Yampa River subwatershed 3	1975–76	3	1	<2	550	400 (85th)	140 (acute)/75 (chronic)
123	WALTON CREEK NEAR STEAMBOAT SPRINGS, CO.	09238500	3	Yampa River subwatershed 2	1984–85	3	1	<1	7	6 (85th)	2.26 (acute)/1.78 (chronic)
	Iron, dissolved (micrograms per liter)										
63	LITTLE MORRISON CK @ RD 18A	12896	3	Yampa River subwatershed 1	[2]2001–07	7	0	130	570	318 (85th)	300 WS (chronic, SMCL)
166	Yampa River Library	CDOWRW-12	2c	Yampa River subwatershed 2	1991–92	4	0	193	1,001	962 (85th)	300 WS (chronic, SMCL)
170	Yampa River Stagecoach Res	CDOWRW-8	2a	Yampa River subwatershed 1	1990–92, 2004	16	0	31	667	651 (85th)	300 WS (chronic, SMCL)
	Iron, total recoverable (micrograms per liter)										
16	CHIMNEY CREEK AT TRAPPER, CO.	400612106524800	5	Yampa River subwatershed 1	1975–76	4	0	740	1,400	1,150 (50th)	1,000 (chronic)
17	COW CR. NR. STEAMBOAT SPRINGS,CO.	402836106550100	3	Yampa River subwatershed 2	1982	4	0	1,500	3,700	2,650 (50th)	1,000 (chronic)
51	FISH CREEK AT MOUTH NEAR MILNER, CO.	402530106585700	13b	Yampa River subwatershed 3	1975–76, 1982	10	0	370	37,000	6,350 (50th)	1,000 (chronic)
117	TROUT CREEK ABOVE MILNER, CO.	402720106591200	13f	Yampa River subwatershed 3	1982	4	0	6,000	24,000	9,350 (50th)	1,000 (chronic)
164	Yampa River East Br	CDOWRW-13	2c	Yampa River subwatershed 3	1998	7	0	11	9,636	2,728 (50th)	1,000 (chronic)

Table 10. Stream water-quality sites with data in exceedance of Colorado Department of Public Health and Environment in-stream water-quality standards for selected physical properties and water-quality constituents and U.S. Environmental Protection Agency recommended concentrations for total phosphorus, Upper Yampa River watershed, Colorado, 1975 through 2009.—Continued

[No., number; Min., minimum; Max., maximum; WS, water supply; SMCL, secondary maximum contaminant level; USEPA, U.S. Environmental Protection Agency; <, less than. Subwatershed definitions: Yampa River subwatershed 1, Yampa River and tributaries upstream from Chuck Lewis State Wildlife Area; Yampa River subwatershed 2, Yampa River and tributaries from Chuck Lewis State Wildlife Area to Elk River confluence; Elk River subwatershed, Elk River and tributaries; Yampa River subwatershed 3, Yampa River and tributaries from Elk River confluence to Town of Hayden; Yampa River subwatershed 4, Yampa River and tributaries from Town of Hayden to Elkhead Creek confluence; Elkhead Creek subwatershed, Elkhead Creek and tributaries. Water-quality standards are from Colorado Department of Public Health and Environment (2009a, 2010); standards are for protection of aquatic life, unless otherwise stated. Recommended concentrations for total phosphorus are from U.S. Environmental Protection Agency (2000). Nonattainment of a standard is based on the 15th, 50th, or 85th percentile value shown in the table. See appendix 3 for additional site information. Descriptions of stream segments are in Colorado Department of Public Health and Environment (2010)]

Site no. (see figures 8, 12)	Site name in Upper Yampa River watershed water-quality database	Site identifier	Stream segment	Subwatershed	Period of water-quality record (calendar year)	No. of samples	No. of censored values[1]	Min. value	Max. value	Percentile value (percentile)	In-stream water-quality standard or recommended value
	Manganese, dissolved (micrograms per liter)										
63	LITTLE MORRISON CK @ RD 18A	12896	3	Yampa River subwatershed 1	[2]2001–07	7	0	29	83	75 (85th)	50 WS (chronic, SMCL)
83	OAK CK D/S TOWN OF OAK CREEK @ CR 27	12892	6	Yampa River subwatershed 1	1999–2007	14	0	14	100	52 (85th)	50 WS (chronic, SMCL)
91	Oak Creek Decker Pk	CDOWRW-80	6	Yampa River subwatershed 1	1991–92	13	0	31	116	81 (85th)	50 WS (chronic, SMCL)
115	TROUT CK NR. MOUTH	12876	13f	Yampa River subwatershed 3	[2]1996–07	22	0	11	75	68 (85th)	50 WS (chronic, SMCL)
122	WALTON CR. NEAR MOUTH @ HWY 40	12834	3	Yampa River subwatershed 2	1999, 2006–07	5	0	16	130	72 (85th)	50 WS (chronic, SMCL)
124	WALT [Walton Creek 10m above Hwy 40 bridge]	WALT	3	Yampa River subwatershed 2	2007–08	4	0	12	133	84 (85th)	50 WS (chronic, SMCL)
129	YAMPA ABOVE OAK CREEK CONFLUENCE	000088	2a	Yampa River subwatershed 1	1988–93	22	4	[6]16	160	78 (85th)	50 WS (chronic, SMCL)
138	YAMPA R. BLW STAGECOACH RES.	12808	2a	Yampa River subwatershed 1	1996–99	14	1	<4	220	180 (85th)	50 WS (chronic, SMCL)
139	YAMPA R. D/S STAGECOACH RES. DAM	12808P	2a	Yampa River subwatershed 1	[2]2000–07	21	0	50	210	170 (85th)	50 WS (chronic, SMCL)
140	YAMPA R. N. OF HAYDEN @ CALIFORNIA PARK RD	12802	2c	Yampa River subwatershed 3	[2]1996–99	13	1	<4	68	58 (85th)	50 WS (chronic, SMCL)
143	YAMPA RIVER AB OAK CREEK NR STEAMBOAT SPGS, CO.	40235610650000	2a	Yampa River subwatershed 1	1975–76	4	1	<10	60	60 (85th)	50 WS (chronic, SMCL)
143	YAMPA R. ABV OAK CREEK	12811	2a	Yampa River subwatershed 1	1996–2002	30	0	5	120	87 (85th)	50 WS (chronic, SMCL)
153	YAMPA RIVER AT STEAMBOAT SPRINGS, CO	09239500	2c	Yampa River subwatershed 2	[2]1975–2009	68	2	[6]3.5	130	76 (85th)	50 WS (chronic, SMCL)
153	YAMPA R. @ 5TH ST. BRIDGE IN STEAMBOAT	12806	2c	Yampa River subwatershed 2	[2]1996–2007	37	1	<4	120	72 (85th)	50 WS (chronic, SMCL)

Table 10. Stream water-quality sites with data in exceedance of Colorado Department of Public Health and Environment in-stream water-quality standards for selected physical properties and water-quality constituents and U.S. Environmental Protection Agency recommended concentrations for total phosphorus, Upper Yampa River watershed, Colorado, 1975 through 2009.—Continued

[No., number; Min., minimum; Max., maximum; WS, water supply; SMCL, secondary maximum contaminant level; USEPA, U.S. Environmental Protection Agency; <, less than. Subwatershed definitions: Yampa River subwatershed 1, Yampa River and tributaries upstream from Chuck Lewis State Wildlife Area; Yampa River subwatershed 2, Yampa River and tributaries from Chuck Lewis State Wildlife Area to Elk River confluence; Elk River subwatershed, Elk River and tributaries; Yampa River subwatershed 3, Yampa River and tributaries from Elk River confluence to Town of Hayden; Yampa River subwatershed 4, Yampa River and tributaries from Town of Hayden to Elkhead Creek confluence; Elkhead Creek subwatershed, Elkhead Creek and tributaries. Water-quality standards are from Colorado Department of Public Health and Environment (2009a, 2010); standards are for protection of aquatic life, unless otherwise stated. Recommended concentrations for total phosphorus are from U.S. Environmental Protection Agency (2000). Nonattainment of a standard is based on the 15th, 50th, or 85th percentile value shown in the table. See appendix 3 for additional site information. Descriptions of stream segments are in Colorado Department of Public Health and Environment (2010)]

Site no. (see figures 8, 12)	Site name in Upper Yampa River watershed water-quality database	Site identifier	Stream segment	Subwatershed	Period of water-quality record (calendar year)	No. of samples	No. of censored values[1]	Min. value	Max. value	Percentile value (percentile)	In-stream water-quality standard or recommended value
					Manganese, dissolved (micrograms per liter)—Continued						
158	Yampa River Below Stagecoach	CDOWRW-81	2a	Yampa River subwatershed 1	2004	5	0	54	216	180 (85th)	50 WS (chronic, SMCL)
166	Yampa River Library	CDOWRW-12	2c	Yampa River subwatershed 2	1991–92	8	0	13	94	86 (85th)	50 WS (chronic, SMCL)
168	YAMPA RIVER NEAR HAYDEN, CO.	09244400	2c	Yampa River subwatershed 3	1979	3	0	20	60	54 (85th)	50 WS (chronic, SMCL)
171	Yampa River SWA Br	CDOWRW-10	2a	Yampa River subwatershed 1	1991–92	15	0	11	144	83 (85th)	50 WS (chronic, SMCL)
					Selenium, dissolved (micrograms per liter)						
22	DRY CK @ HAYDEN	12852	13d	Yampa River subwatershed 4	[2]2001–05	7	2	<1	120	52 (85th)	18.4 (acute)/4.6 (chronic)
59	GRASSY CREEK NEAR MOUNT HARRIS, CO.	09244300	13e	Yampa River subwatershed 3	1975–76, 1982	5	3	<1	34	25 (85th)	18.4 (acute)/4.6 (chronic)
59	GRASSY CK @ RD. 27A	12853	13e	Yampa River subwatershed 3	2001, 2004–07	7	1	<1	42	26 (85th)	18.4 (acute)/4.6 (chronic)
98	SAGE CK @ RD. 27	12851	13e	Yampa River subwatershed 3	2001, 2004–07	7	1	<1	290	188 (85th)	18.4 (acute)/4.6 (chronic)
101	SAGE CREEK NEAR HAYDEN, CO.	402918107094400	13e	Yampa River subwatershed 3	1975–76	5	1	<1	52	24 (85th)	18.4 (acute)/4.6 (chronic)
112	STOKES GULCH NEAR HAYDEN, CO.	09244470	13d	Yampa River subwatershed 4	1978–81	5	0	24	300	166 (85th)	18.4 (acute)/4.6 (chronic)
158	Yampa River Below Stagecoach	CDOWRW-81	2a	Yampa River subwatershed 1	2004	3	0	2.2	7.9	6.5 (85th)	18.4 (acute)/4.6 (chronic)

[1]Censored values can be expressed as values less than the laboratory reporting level.

[2]Sample collection did not occur during every year in the period of record.

[3]Continuous monitoring (15-minute intervals) of water temperature only occurred at Yampa River at Steamboat Springs (site 153). Temperature measurements were collected on most sample days from July 26, 2002, through April 13, 2005. See figure 9 and table 11 for temperature standards and the number of exceedances of the standards.

[4]For stream segments with dissolved solids data, the standard has being applied to stream segments that have standards for unfiltered sulfate and chloride.

[5]For stream segments with unfiltered total phosphorus data, the recommendation has been applied to stream segments that have a standard for unfiltered nitrate.

[6]Minimum censored value is greater than minimum detected value.

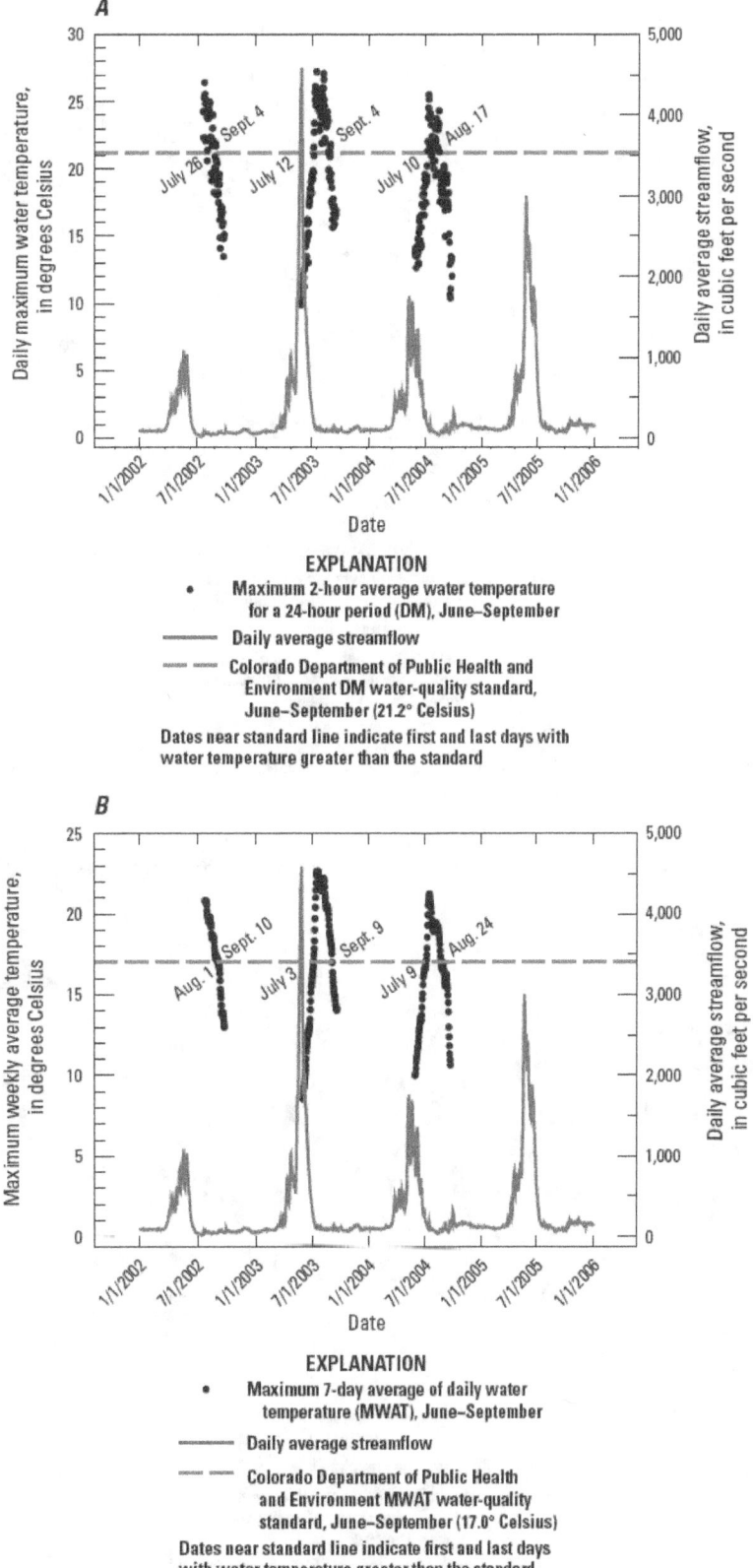

Figure 9. Daily Maximum Temperature (DM), Maximum Weekly Average Temperature (MWAT), and daily average streamflow with exceedances of Colorado Department of Public Health and Environment (CDPHE) in-stream water-quality standards for water temperature for (*A*) June–September DM, (*B*) June–September MWAT, (*C*) October–May DM, and (*D*) October–May MWAT for Yampa River at Steamboat Springs (site 153), Upper Yampa River watershed, Colorado, 2002 though 2005.

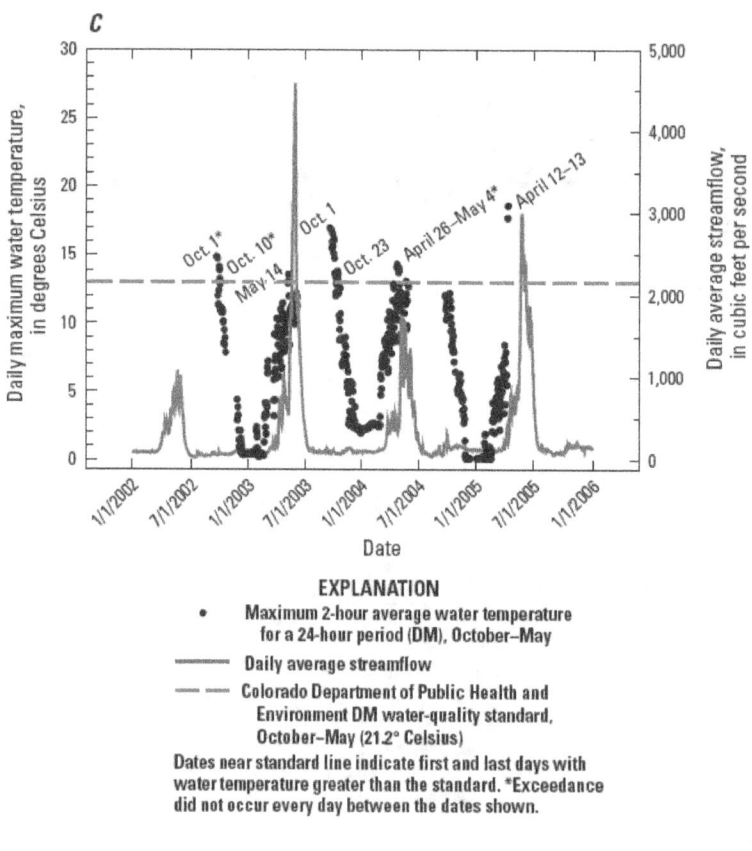

EXPLANATION
• Maximum 2-hour average water temperature
  for a 24-hour period (DM), October–May
— Daily average streamflow
— — Colorado Department of Public Health and
  Environment DM water-quality standard,
  October–May (21.2° Celsius)
Dates near standard line indicate first and last days with
water temperature greater than the standard. *Exceedance
did not occur every day between the dates shown.

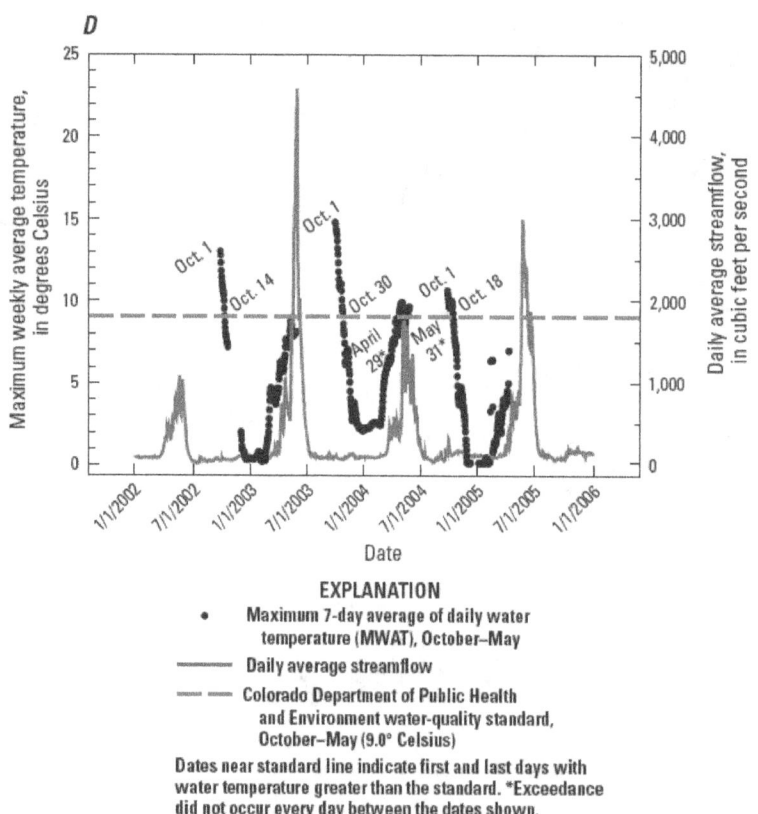

EXPLANATION
• Maximum 7-day average of daily water
  temperature (MWAT), October–May
— Daily average streamflow
— — Colorado Department of Public Health
  and Environment water-quality standard,
  October–May (9.0° Celsius)
Dates near standard line indicate first and last days with
water temperature greater than the standard. *Exceedance
did not occur every day between the dates shown.

Figure 9. Daily Maximum Temperature (DM), Maximum Weekly Average Temperature (MWAT), and daily average streamflow with exceedances of Colorado Department of Public Health and Environment (CDPHE) in-stream water-quality standards for water temperature for (*A*) June–September DM, (*B*) June–September MWAT, (*C*) October–May DM, and (*D*) October–May MWAT for Yampa River at Steamboat Springs (site 153), Upper Yampa River watershed, Colorado, 2002 though 2005.—Continued

Table 11. Number of sample days with water temperature at Yampa River at Steamboat Springs (site 153) not meeting Colorado Department of Public Health and Environment in-stream water-quality standards, July 26, 2002, through April 13, 2005.

[C, degrees Celsius; DM, Daily Maximum Temperature; MWAT, Maximum Weekly Average Temperature. Water-quality standards are from Colorado Department of Public Health and Environment (2009a, 2010). Continuous (15-minute interval) temperature data were available for July 26, 2002 through April 13, 2005. See figure 3 for location of site]

In-stream water-quality standard (°C)	Number of sample days	Number of sample days not meeting standard	Percentage of sample days not meeting standard (percent)
21.2 (cold, acute DM, June–September)	311	113	36
17.0 (cold, chronic MWAT, June–September)	305	151	50
13.0 (cold, acute DM, October–May)	578	32	5.5
9.0 (cold, chronic MWAT, October–May)	608	88	14

Dissolved oxygen in the UYRW was measured for 2,797 samples from 1975 through 2009 and ranged from 1.2 to 17.8 mg/L with a median of 9.8 mg/L (table 8). Median dissolved oxygen concentrations were between 9.8 and 10.1 mg/L for each subwatershed, except Yampa River subwatershed 4. In this subwatershed, the median dissolved oxygen concentration was lower (9.2 mg/L) and the median water temperature was higher than those in any other subwatershed in the watershed. The lowest dissolved oxygen concentrations (less than 5.0 mg/L) in the watershed generally occurred during late June through August when streamflow was lower and water temperatures were higher than at other times of the year. The CDPHE standard for dissolved oxygen was met at about 99 percent of sites with data available for comparison to the standard, indicating adequate conditions for aquatic life. Two sites in this study, Little White Snake Creek (site 66, stream not named on fig. 3) and Martin Creek (site 70, stream not named on fig. 3), had concentrations that were not in attainment of the CDPHE standard (table 10). Little White Snake Creek is on the CDPHE 2012 monitoring and evaluation list for dissolved oxygen (Colorado Department of Public Health and Environment, 2012). Data for this site were available for 2001, 2006, and 2007. Determination of nonattainment of the standard for Martin Creek was based on data collected from 1986 through 1988.

Hardness generally is measured by the presence of the cations calcium and magnesium in water and is reported in terms of an equivalent concentration of calcium carbonate. Hardness can affect the anthropogenic uses of water and the toxicity of metals to aquatic life. Metal toxicity can increase as hardness decreases (Santore and others, 2001). Waters draining igneous rocks generally contain little hardness because of the absence of calcium and magnesium cations in the water. Waters draining sandstones and shales, as well as other sedimentary rocks that contain carbonate, have harder water. Additionally, additives from municipal water treatment, agricultural fertilizers, and applications of chemicals for winter road maintenance can increase hardness in water.

In the UYRW, hardness was measured for 2,434 samples from 1975 through 2009 and ranged from 4 to 4,000 mg/L with a median of 162 mg/L (table 8). Hardness generally was lowest (softer water) in the Elk River subwatershed and Yampa River subwatershed 2, where igneous and metamorphic rocks underlie much of the land surface, and highest (harder water) in Yampa River subwatersheds 3 and 4 (table 8), where sedimentary rocks predominate. Hardness was generally higher when streamflow was primarily base flow from groundwater than during snowmelt runoff.

Acid neutralizing capacity (determined on an unfiltered water sample) and alkalinity (determined on a filtered sample) measure the ability of a water sample to neutralize inputs of acid from precipitation, wastewater, or mine drainage (Rounds, 2006). Bicarbonate and carbonate are the main buffering materials. Water with low ANC is susceptible to pH change with the addition of acidic water; water with high ANC is buffered and resists pH change. Waters with high ANC occur in areas with sedimentary rocks and carbonate-rich materials.

In the UYRW, ANC was measured for 2,244 samples collected at 134 sites, most commonly for Yampa River subwatersheds 1 and 3. Values for ANC ranged from 2 to 660 mg/L with a median of 136 mg/L (table 8). Lower values typically occurred during snowmelt runoff than others times of the year. As with hardness, the Elk River subwatershed and Yampa River subwatershed 2 had the lowest median ANC, and the Yampa River subwatersheds 3 and 4 had the highest median ANC because of the underlying geology (figs. 5, 6C). In the Rocky Mountains, contaminants in precipitation, including sulfuric and nitric acids, pose a threat to alpine and subalpine ecosystems, such as Mount Zirkel and Flat Tops Wilderness areas in the UYRW (fig. 1), because water, local soil types, and bedrock have little capacity to buffer acidic inputs (Turk and Spahr, 1991). As the accumulated snowpack melts, a pulse of strong acids can be released into the aquatic system (Campbell and Turk, 1989).

Dissolved Solids and Major Ions

In most natural waters, dissolved solids are composed of commonly occurring major ions such as calcium, magnesium, sodium, potassium, bicarbonate, carbonate, sulfate, chloride, fluoride, and silica. For this study, dissolved solids are materials dissolved in water that pass through a 0.45-micron filter. Most dissolved ions originate when water reacts with mineral assemblages in rocks and soils near the land surface (Hem, 1992). The amount and composition of dissolved solids can be used to identify sources of and changes in water chemistry at different times of the year.

In the UYRW, dissolved solids were measured in 1,743 samples collected from 1975 through 2009 (tables 6, 7). Concentrations ranged from 6.5 to 9,280 mg/L, and the median was 216 mg/L (table 12). Among the subwatersheds, the median concentration was highest for Yampa River subwatershed 4, which likely resulted from the weathering of sedimentary rocks that underlie the subwatershed, and the lowest for the Elk River subwatershed, an area that drains relatively non-reactive igneous and metamorphic rocks. Because values of dissolved solids and specific conductance in water are proportional, the spatial distribution of dissolved solids concentrations in the UYRW is similar to that of specific conductance (fig. 6A). Tributaries to the Yampa River in subwatershed 2 and the Elk River, the major tributary to the Yampa River in Yampa River subwatershed 3, carried fewer dissolved solids than tributaries in Yampa River subwatersheds 1 and 4. The 85th percentile concentration of dissolved solids for one site, Trout Creek near the mouth (site 115), in Yampa River subwatershed 3 was greater than the nonenforceable SMCL of 500 mg/L (table 10). High dissolved solids concentrations can limit the use of water for municipal and agricultural uses and can affect aquatic life.

Major ions in water are a product of the interaction of groundwater and aquifer materials, lithology, soils, and human activity. In this section, major ions are dissolved concentrations unless stated otherwise. A total of 1,921 samples analyzed for major ions were collected in the UYRW from 1975 through 2009 at 130 stream sites (tables 6, 7). The most sites sampled and the most samples collected were in Yampa River subwatershed 3; the fewest sites and fewest samples were recorded for the Elkhead Creek subwatershed and Yampa River subwatershed 4, respectively (tables 6, 7).

A total of 172 samples had sufficient data for calcium, magnesium, potassium, sodium, bicarbonate, chloride, fluoride, and sulfate to determine the ionic composition of water. For most subwatersheds, the primary cations in solution were calcium or calcium plus magnesium, and the primary anions were bicarbonate or bicarbonate plus sulfate (fig. 10). The most diverse water type was for streams in Yampa River subwatershed 4. Hubberson and Watering Trough Gulches (not named in figure 3) near Hayden (sites 61 and 125, respectively), which are underlain by the Williams Fork Formation, had a calcium plus magnesium and bicarbonate plus sulfate water type. Stokes Gulch (not named in figure 3) near Hayden

(site 112), underlain by the Lewis Shale, had a sodium plus potassium plus magnesium and sulfate water type. In the Williams Fork aquifer of the Mesaverde Group, the dissolution of calcite and dolomite from limey shales, limestones, and dolomitic limestones is a source of calcium and magnesium (Robson and Stewart, 1990). In areas with marine shales, such as the Lewis Shale, the exchange of calcium and magnesium ions in solution with sodium ions on the clay minerals in sodium-rich marine shales is the principal source of sodium in the groundwater (Robson and Stewart, 1990).

Median dissolved sulfate concentrations were highest in Yampa River subwatersheds 3 and 4 (fig. 6D, table 12) because of the prevalence of sedimentary rocks in the subwatersheds. Attainment of the CDPHE water-supply standard for unfiltered sulfate was not met at about 4 percent (2 of 52) of the sites sampled. One site not in attainment was on Little White Snake Creek (site 66) in Yampa River subwatershed 1, and the other site was on Trout Creek (site 115) in Yampa River subwatershed 3 (fig. 8, table 10). Both sites have drainage basins that overlie sedimentary rocks. The CDPHE has not established a sulfate standard for many stream segments in Yampa River subwatershed 3 and a few segments in the Yampa River subwatersheds 1 and 4 because of naturally high concentrations of sulfate.

All unfiltered chloride concentrations in the UYRW were 76 mg/L or less. The concentrations were well below the CDPHE water-quality standard of 250 mg/L (table 12).

Concentrations of major ions and dissolved solids typically were lower during snowmelt runoff in May and June than at other times of the year, as illustrated by monthly dissolved sulfate concentrations for Yampa River at Milner (site 151) (fig. 7B). This seasonal variation in concentrations is primarily because of the water source and increased volume of streamflow. Snowmelt typically is lower in dissolved materials than groundwater inflow to streams because snowmelt runoff has little soil and (or) rock interaction compared to groundwater. Because of the strong influence of snowmelt runoff on water quality, it is difficult to identify other natural and human factors that may affect major ion and dissolved solids concentrations. Typically, major ion and dissolved solids concentrations increase steadily through the summer and fall as inflow from groundwater becomes the major source of water to a stream.

Only one site, Yampa River at Steamboat Springs (site 153), had sufficient dissolved solids and major ion data for temporal trend analysis. A statistically significant (p-value 0.03) downward trend in dissolved solids concentrations was identified for data collected from 1997 through 2008 (table 9). The rate of changed was small, about 3 percent per year (magnitude about 3.9 mg/L per year). The median dissolved solids concentration at the site was 153 mg/L. A similar downward trend was identified for specific conductance at the site. No statistically significant (p-values 0.06–0.77) trends were identified for calcium, magnesium, sodium, potassium, sulfate, chloride, and silica concentrations in stream water at Yampa River at Steamboat Springs (site 153).

Table 12. Summary statistics for dissolved solids and selected major ion water-quality data and Colorado Department of Public Health and Environment in-stream water-quality standards for stream sampling sites in the Upper Yampa River watershed and subwatersheds, Colorado, 1975 through 2009.

[No., number; WS, water supply; SMCL, secondary maximum contaminant level; --, no water-quality standard; <, less than. All constituent values are reported in milligrams per liter. Number of significant figures for individual constituents may vary because data are from different sources and analytical periods. Water-quality standards are from Colorado Department of Public Health and Environment (2009a, 2010). See table 10 for sites in the Upper Yampa River watershed and subwatersheds with data for dissolved solids and major ions not in attainment of in-stream water-quality standards. Descriptions of stream segments are in Colorado Department of Public Health and Environment (2010)]

Constituent	No. of sites	No. of samples	No. of censored values[1]	Minimum value[2]	Median value	Maximum value	In-stream water-quality standard	No. of sites with data not in attainment of standard
				Upper Yampa River watershed				
Dissolved solids	110	1,743	1	6.5	216	9,280	[3,4]500 WS (SMCL) or none	1
Calcium, dissolved	75	1,064	0	1.4	39.5	480	--	--
Magnesium, dissolved	75	1,066	0	0.2	15.1	810	--	--
Sodium, dissolved	65	1,022	0	0.6	18	1,400	--	--
Potassium, dissolved	65	1,019	0	0.2	2.2	19	--	--
Bicarbonate, dissolved	28	184	0	14	147	550	--	--
Sulfate, dissolved	69	1,020	16	0.65	64.6	6,200	--	--
Sulfate, unfiltered	52	833	67	<3	36	4,300	[4]250 WS (SMCL) or none	2
Chloride, dissolved	66	1,016	18	<0.1	3.9	240	--	--
Chloride, unfiltered	16	101	5	<0.002	6	76	[4]250 WS (SMCL) or none	0
Fluoride, dissolved	68	992	172	<0.001	0.2	1.0	--	--
Silica, dissolved	66	1,012	0	0.009	8.8	28	--	--
			Yampa River and tributaries upstream from Chuck Lewis State Wildlife Area					
			(Yampa River subwatershed 1; stream segments 2a, 2c, 3, 4, 5, 6, 7)					
Dissolved solids	28	429	0	43	260	1,130	[3,4]500 WS (SMCL) or none	0
Calcium, dissolved	8	84	0	21	53.7	130	--	--
Magnesium, dissolved	8	84	0	7.1	19.7	36	--	--
Sodium, dissolved	6	74	0	3.5	12	32	--	--
Potassium, dissolved	6	74	0	0.4	2.2	4.5	--	--
Bicarbonate, dissolved	3	7	0	160	211	257	--	--
Sulfate, dissolved	6	73	6	<5	55	250	--	--
Sulfate, unfiltered	22	389	3	<3	45	340	[4]250 WS (SMCL) or none	1
Chloride, dissolved	6	74	0	0.4	2.4	14	--	--
Chloride, unfiltered	10	46	2	<1	4	76	[4]250 WS (SMCL) or none	0
Fluoride, dissolved	6	74	5	<0.001	0.2	0.6	--	--
Silica, dissolved	6	74	0	6.9	19	28	--	--
			Yampa River and tributaries from Chuck Lewis State Wildlife Area to Elk River confluence					
			(Yampa River subwatershed 2; stream segments 2c, 3, 20a)					
Dissolved solids	13	166	1	6.5	89.4	471	[3]500 WS (SMCL)	0
Calcium, dissolved	19	157	0	1.4	12.1	100	--	--
Magnesium, dissolved	19	157	0	0.2	2.7	18	--	--
Sodium, dissolved	11	125	0	0.7	2.6	30	--	--
Potassium, dissolved	11	125	0	0.2	0.9	4.9	--	--
Bicarbonate, dissolved	3	26	0	15	111	157	--	--
Sulfate, dissolved	12	125	4	0.65	5.0	160	--	--
Sulfate, unfiltered	6	61	17	<3	14	54	250 WS (SMCL)	0
Chloride, dissolved	12	125	3	<0.1	1.1	12	--	--
Fluoride, dissolved	12	123	61	0.03	0.08	0.9	--	--
Silica, dissolved	12	120	0	2.4	8.9	18	--	--
			Elk River and tributaries					
			(Elk River subwatershed; stream segments 8, 20a)					
Dissolved solids	17	200	0	12.7	29.6	180	[3]500 WS (SMCL)	0
Calcium, dissolved	10	130	0	1.9	5.9	29	--	--
Magnesium, dissolved	10	130	0	0.3	1.2	10.6	--	--
Sodium, dissolved	10	130	0	0.6	1.6	9.2	--	--
Potassium, dissolved	10	130	0	0.2	0.8	4.7	--	--
Bicarbonate, dissolved	3	10	0	14	38.5	78	--	--
Sulfate, dissolved	10	128	0	0.73	3.2	53	--	--
Sulfate, unfiltered	7	90	38	<3	7	50	250 WS (SMCL)	0
Chloride, dissolved	10	128	15	<0.1	0.3	4.6	--	--
Chloride, unfiltered	1	7	1	2	4	4	250 WS (SMCL)	0
Fluoride, dissolved	9	101	60	<0.1	<0.1	0.4	--	--
Silica, dissolved	10	129	0	1.7	7.0	14.7	--	--

Table 12.   Summary statistics for dissolved solids and selected major ion water-quality data and Colorado Department of Public Health and Environment in-stream water-quality standards for stream sampling sites in the Upper Yampa River watershed and subwatersheds, Colorado, 1975 through 2009.—Continued

[No., number; WS, water supply; SMCL, secondary maximum contaminant level; --, no water-quality standard; <, less than. All constituent values are reported in milligrams per liter. Number of significant figures for individual constituents may vary because data are from different sources and analytical periods. Water-quality standards are from Colorado Department of Public Health and Environment (2009a, 2010). See table 10 for sites in the Upper Yampa River watershed and subwatersheds with data for dissolved solids and major ions not in attainment of in-stream water-quality standards. Descriptions of stream segments are in Colorado Department of Public Health and Environment (2010)]

Constituent	No. of sites	No. of samples	No. of censored values[1]	Minimum value[2]	Median value	Maximum value	In-stream water-quality standard	No. of sites with data not in attainment of standard
Yampa River and tributaries from Elk River confluence to Town of Hayden								
(Yampa River subwatershed 3; stream segments 2c, 11, 12, 13a, 13b, 13c, 13e, 13f, 20a)								
Dissolved solids	33	647	0	10	321	5,650	[3,4]500 WS (SMCL) or none	1
Calcium, dissolved	23	429	0	6.6	78	480	--	--
Magnesium, dissolved	23	431	0	1.5	37	310	--	--
Sodium, dissolved	23	429	0	1.9	32	540	--	--
Potassium, dissolved	23	425	0	0.5	3.4	19	--	--
Bicarbonate, dissolved	11	116	0	21	160	510	--	--
Sulfate, dissolved	26	429	6	1.8	170	2,300	--	--
Sulfate, unfiltered	13	248	9	<3	30	4,300	[4]250 WS (SMCL) or none	1
Chloride, dissolved	23	424	0	0.3	6.5	59	--	--
Chloride, unfiltered	4	47	2	<0.002	12	53	[4]250 WS (SMCL) or none	0
Fluoride, dissolved	26	430	10	<0.001	0.2	1.0	--	--
Silica, dissolved	23	423	0	0.2	8	18	--	--
Yampa River and tributaries from Town of Hayden to Elkhead Creek confluence								
(Yampa River subwatershed 4; stream segments 2c, 12, 13d)								
Dissolved solids	10	101	0	60	709	9,280	[3,4]500 WS (SMCL) or none	0
Calcium, dissolved	8	78	0	33	110	280	--	--
Magnesium, dissolved	8	78	0	13	72	810	--	--
Sodium, dissolved	8	78	0	11	54	1,400	--	--
Potassium, dissolved	8	79	0	2.1	5	12	--	--
Bicarbonate, dissolved	4	17	0	144	360	550	--	--
Sulfate, dissolved	8	80	0	46	360	6,200	--	--
Sulfate, unfiltered	3	26	0	8	57	3,800	[4]250 WS (SMCL) or none	0
Chloride, dissolved	8	80	0	3.4	12.5	240	--	--
Chloride, unfiltered	1	1	0	13	13	13	[4]250 WS (SMCL) or none	0
Fluoride, dissolved	9	81	1	<0.001	0.3	0.6	--	--
Silica, dissolved	8	80	0	0.1	8.1	19	--	--
Elkhead Creek and tributaries								
(Elkhead Creek subwatershed; stream segments 14, 15, 20b)								
Dissolved solids	9	200	0	68	202	635	[3]500 WS (SMCL)	0
Calcium, dissolved	7	186	0	13	31	71	--	--
Magnesium, dissolved	7	186	0	3.8	12.3	39.3	--	--
Sodium, dissolved	7	186	0	3.6	19.0	77.5	--	--
Potassium, dissolved	7	186	0	0.76	1.6	4.5	--	--
Bicarbonate, dissolved	4	8	0	72	167	232	--	--
Sulfate, dissolved	7	185	0	11.3	52.2	347	--	--
Sulfate, unfiltered	1	19	0	17	82	130	250 WS (SMCL)	0
Chloride, dissolved	7	185	0	0.3	3.0	12	250 WS (SMCL)	0
Fluoride, dissolved	6	183	35	0.081	0.13	0.3	--	--
Silica, dissolved	7	186	0	0.009	9.9	16.1	--	--

[1]Censored values can be expressed as values less than the laboratory reporting level.

[2]For some constituents with censored data, the minimum censored value is greater than the minimum detected value that is shown.

[3]For stream segments with dissolved solids data, the standard has been applied to stream segments that have standards for unfiltered sulfate and chloride.

[4]Water-quality standard varies by stream segment. See Colorado Department of Public Health and Environment (2010).

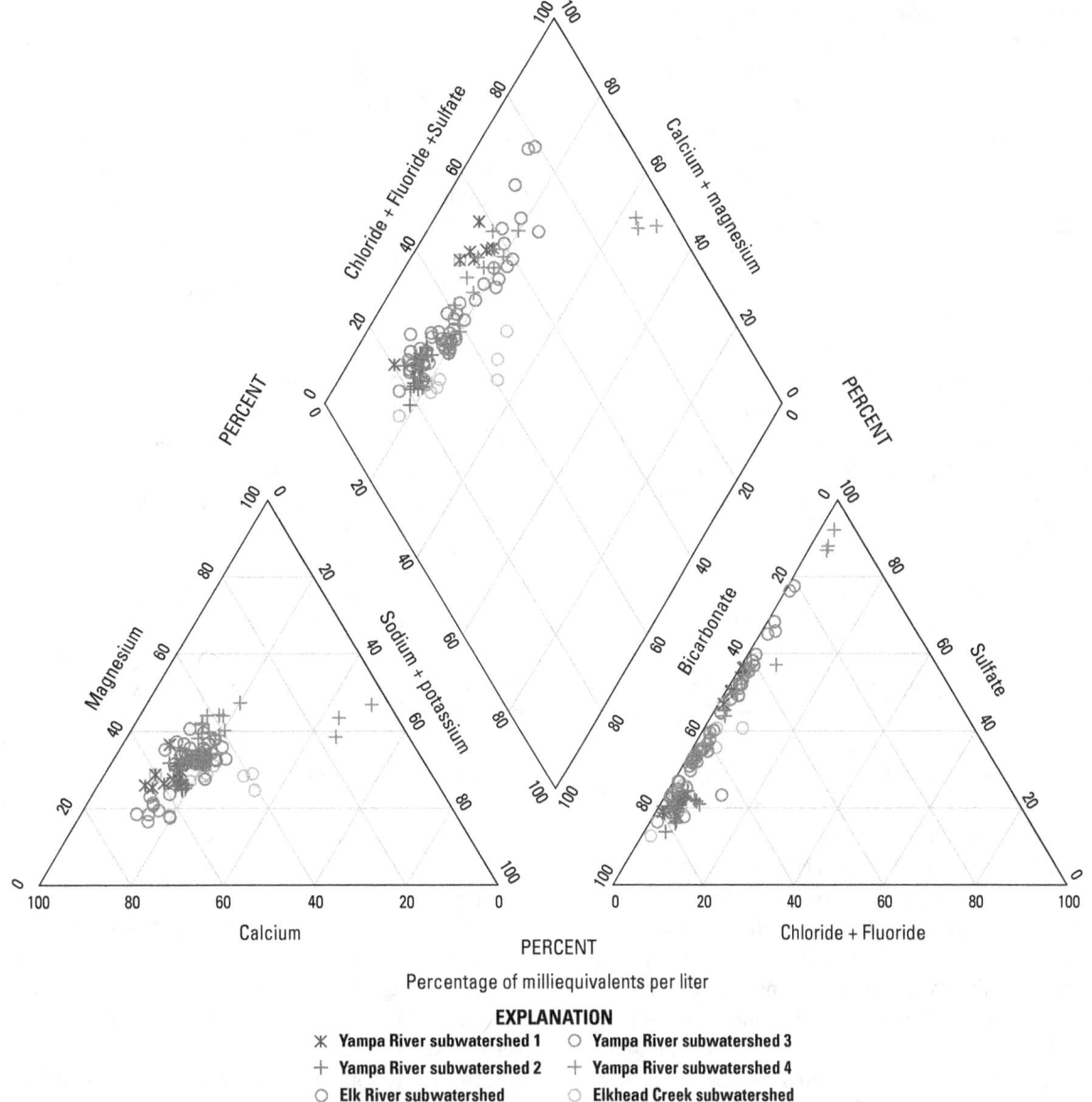

Percentage of milliequivalents per liter

EXPLANATION

✳ Yampa River subwatershed 1	○ Yampa River subwatershed 3
+ Yampa River subwatershed 2	+ Yampa River subwatershed 4
○ Elk River subwatershed	○ Elkhead Creek subwatershed

Figure 10.   Major cation and anion percentages and water type for water-quality samples collected from selected stream sites, by subwatershed, Upper Yampa River watershed, Colorado, 1975 through 2009.

Nutrients

Nutrients (nitrite, nitrate, ammonia, phosphorus, and orthophosphate) in stream water provide essential food for plants and animals. They can occur naturally in water from the weathering and erosion of rocks and soils, breakdown of organic matter, and atmospheric deposition but also can result from human activities. These activities can include application of fertilizers, runoff from agricultural and urban areas, soil erosion, effluent from the wastewater-treatment process, seepage from septic tanks, detergents, animal waste, and combustion of fuels such as coal, petroleum, and wood. Nitrite typically is found in low concentrations in streams, often at concentrations near laboratory detection levels, because it

is unstable in aerated water. High levels of nitrite generally indicate pollution through disposal of sewage or organic waste (Hem, 1992). Nitrate is a more stable species of nitrogen in aerated water. It can be found in low concentrations in streams and lakes because it is readily consumed by aquatic plants but can also occur at high concentrations because of natural sources or human activities. Excessive concentrations of nitrate in drinking water may cause methemoglobinemia, commonly known as blue baby syndrome, in small children (Hem, 1992). Ammonia occurs in water as ammonium (NH_4+) and (or) un-ionized ammonia (NH_3). Ammonium is the form of ammonia that is the available nutrient; un-ionized ammonia is the form that can be toxic to fish in excessive concentrations. Toxicity varies depending on species and is pH and temperature dependent (toxicity increases as pH and temperature increase), with pH having a larger effect on toxicity that temperature. Phosphorous often adsorbs to the surface of sediment and organic particles, which reduces concentrations of phosphorus in streams. The presence of phosphorus in surface water can indicate that erosion and sediment transport are occurring (Mueller and others, 1995).

An increase in the supply of nutrients, especially nitrates and phosphates, to surface water from natural and (or) artificial sources can result in eutrophication. In eutrophic water, high concentrations of nutrients can stimulate excessive algal growth. As algae die and decompose, dissolved oxygen is consumed, and the amount of dissolved oxygen available for aquatic life is reduced or depleted. Eutrophication in surface water can result in habitat loss, toxic algal blooms, reduction in biodiversity, taste and odor issues, pH fluctuations, and clogged municipal water intake pipes (Wetzel, 1983; Murphy, 2007a).

In the UYRW, 2,162 samples with nutrient (nitrogen and phosphorus) data were collected from 164 stream sites (table 6). The spatial and temporal distribution of sample collection was similar to that for physical properties and major ions. More sites were sampled and more samples were collected in Yampa River subwatersheds 1 and 3 than in other subwatersheds; the fewest sites sampled and fewest samples collected were in the Elkhead Creek subwatershed and Yampa River subwatershed 4, respectively (table 6).

Dissolved nitrite was measured for 628 samples collected from 1981 through 2009; concentrations ranged from less than 0.001 to 0.39 mg/L as nitrogen (N) with a median of 0.002 mg/L as N (all nitrite, nitrate, and total ammonia data that are discussed are in the nitrogen form, as N) (table 13). Concentrations of dissolved nitrite in 85 percent of all samples, and in every sample collected from 2002 through 2009, were 0.01 mg/L or less. Although 9 of 628 individual samples had concentrations of unfiltered nitrite greater than the CDPHE aquatic-life standard of 0.05 mg/L, attainment of the standard was met for all sites. The 85th percentile concentration for each site with three or more samples of unfiltered nitrite data was less than or equal to 0.02 mg/L. Temporal trends testing could not be performed using nitrite data because more than 10 percent of the nitrite concentrations were censored and did not meet the statistical requirements for trends testing.

Dissolved nitrate was measured for 1,096 samples collected from 1975 through 2009. Concentrations ranged from less than 0.005 to 90 mg/L with a median of 0.07 mg/L (table 13). For many sites, more than 50 percent of the data were censored. All samples collected after 1988 had dissolved nitrate concentrations of 1.5 mg/L or less. Median concentrations were highest in the Yampa River subwatersheds 3 and 4 (table 13). A total of 37 samples collected in these two

subwatersheds before 1989 were the only UYRW samples with dissolved nitrate concentrations greater than 5 mg/L. The 85th percentile concentration of unfiltered nitrate data for each site in the UYRW with three of more samples was in attainment of the CDPHE MCL for unfiltered nitrate of 10 mg/L (appendix 2). Attainment also was met for sites with an the agricultural-use standard of 100 mg/L. Statistical requirements for trends testing of dissolved and unfiltered nitrate data were not met. More than 10 percent of the dissolved data were censored, and there were an insufficient number of data values for unfiltered nitrite.

Unfiltered total ammonia was analyzed for 1,027 samples collected from 1975 through 2009. Concentrations ranged from less than 0.01 to 2.3 mg/L with a median of 0.02 mg/L (table 13). More than 60 percent of the data were censored values. Higher concentrations (greater than 0.2 mg/L) were detected in only 10 percent of the samples, and most (85 percent) of these were collected before 1996. From 1996 through 2009, about 50 percent of the concentrations greater than 0.1 mg/L were detected in the Yampa River just downstream from Stagecoach Reservoir (sites 138, 158), most commonly during October through February. The maximum unfiltered total ammonia concentration of 2.3 mg/L also was detected at this location. Denitrification processes occurring within the reservoir could be contributing to the increase in ammonia downstream from the reservoir outlet. Unfiltered total ammonia concentrations in four samples, all collected during 1975 or 1976 at three sites, exceeded the calculated CDPHE aquatic-life standard for each sample, which was 0.48 or 0.68 mg/L. No comparison was made to the 85th percentile concentration for a site, because the standard for a sample varies depending on the pH and water temperature of the sample. Unfiltered total ammonia data could not be tested for temporal trends because more than 10 percent of the data were censored.

Unfiltered total phosphorous was measured for 1,581 samples collected at 144 sites. Only about 15 percent of samples had concentrations that were less than detection levels. Detected concentrations ranged from 0.003 to 3.9 mg/L, and the median was 0.044 mg/L (table 13). For the subwatersheds, the median concentration was lowest (0.009 mg/L) for the Elk River subwatershed and highest (0.06 mg/L or more) for Yampa River subwatersheds 1, 3 and 4. Concentrations greater than 1.0 mg/L were measured in samples from Yampa River subwatersheds 3 and 4, which could result from naturally high concentrations in some sedimentary rocks. The Wadge and Wolf Creek coal beds of the Mesaverde Group in the UYRW portion of the Yampa coal field (primarily in Yampa River subwatershed 3 and 4) have very high contents of phosphorus (30 times greater than the average value for Cretaceous-age coal) because of ash deposits (Affolter, 2000). A seasonal pattern in unfiltered total phosphorus concentrations was evident

Table 13. Summary statistics for selected nutrient water-quality data and Colorado Department of Public Health and Environment in-stream water-quality standards for stream sampling sites in the Upper Yampa River watershed and subwatersheds, Colorado, 1975 through 2009.

[No., number; mg/L, milligrams per liter; N, nitrogen; <, less than; --, no water-quality standard; MCL, maximum contaminant level; TVS, table value standard; P, phosphorus; E, estimated; nc, not computed. Number of significant figures for individual constituents may vary because data are from different sources and analytical periods. Water-quality standards are from Colorado Department of Public Health and Environment (2009a, 2010); standards are for aquatic-life protection, unless otherwise stated. See table 10 for sites in the Upper Yampa River watershed and subwatersheds with data for nutrients not in attainment of in-stream water-quality standards. Descriptions of stream segments are in Colorado Department of Public Health and Environment (2010). Collection of samples for nitrite analysis began in 1981]

Constituent (reporting units)	No. of sites	No. of samples	No. of censored values[1]	Minimum value[2]	Median value	Maximum value	In-stream water-quality standard	No. of sites with data not in attainment of standard
Upper Yampa River watershed								
Nitrite, dissolved (mg/L as N)	42	628	457	<0.001	0.002	0.39	--	--
Nitrite, unfiltered (mg/L as N)	9	95	36	<0.0009	0.002	0.4	0.05	0
Nitrate, dissolved (mg/L as N)	93	1,096	360	<0.005	0.07	90	--	--
Nitrate, unfiltered (mg/L as N)	120	1,217	701	0.007	0.06	48	[3]10 (MCL), 100 (agriculture)	0
Total ammonia, unfiltered (mg/L as N)	87	1,027	635	<0.01	0.02	2.3	[3]TVS or none	0
Total phosphorus, unfiltered (mg/L as P)	144	1,581	243	E 0.003	0.044	3.9	[4]0.05, 0.1	[5,6]18
Orthophosphate, dissolved (mg/L as P)	53	854	367	<0.001	0.01	0.22	--	--
Yampa River and tributaries upstream from Chuck Lewis State Wildlife Area (Yampa River subwatershed 1; stream segments 2a, 2c, 3, 4, 5, 6, 7)								
Nitrite, dissolved (mg/L as N)	2	101	83	<0.01	nc	0.04	--	--
Nitrite, unfiltered (mg/L as N)	4	41	16	<0.0009	0.002	0.04	0.05	0
Nitrate, dissolved (mg/L as N)	11	144	83	0.01	0.04	5	--	--
Nitrate, unfiltered (mg/L as N)	33	477	330	0.007	0.08	1.4	[3]10 (MCL), 100 (agriculture)	0
Total ammonia, unfiltered (mg/L as N)	23	406	239	<0.01	0.03	2.3	[3]TVS or none	0
Total phosphorus, unfiltered (mg/L as P)	33	467	51	<0.005	0.07	0.75	[4]0.05, 0.1	[5,7]8
Orthophosphate, dissolved (mg/L as P)	5	107	25	<0.01	0.02	0.21	--	--
Yampa River and tributaries from Chuck Lewis State Wildlife Area to Elk River confluence (Yampa River subwatershed 2; stream segments 2c, 3, 20a)								
Nitrite, dissolved (mg/L as N)	17	130	101	<0.001	0.002	0.02	--	--
Nitrite, unfiltered (mg/L as N)	1	9	5	<0.01	<0.01	0.02	0.05	0
Nitrate, dissolved (mg/L as N)	26	180	89	<0.005	0.02	0.9	--	--
Nitrate, unfiltered (mg/L as N)	27	155	66	0.01	0.08	0.68	10 (MCL)	0
Total ammonia, unfiltered (mg/L as N)	22	116	70	<0.01	0.01	1.9	TVS	0
Total phosphorus, unfiltered (mg/L as P)	37	242	28	<0.005	0.034	0.921	[4]0.05, 0.1	[5]5
Orthophosphate, dissolved (mg/L as P)	10	107	45	<0.001	0.009	0.07	--	--
Elk River and tributaries (Elk River subwatershed; stream segments 8, 20a)								
Nitrite, dissolved (mg/L as N)	8	93	63	<0.001	<0.001	0.03	--	--
Nitrite, unfiltered (mg/L as N)	2	9	6	0.001	<0.01	0.01	0.05	0
Nitrate, dissolved (mg/L as N)	13	121	23	<0.005	0.05	0.95	--	--
Nitrate, unfiltered (mg/L as N)	15	118	80	0.01	0.05	0.88	10 (MCL)	0
Total ammonia, unfiltered (mg/L as N)	9	98	78	<0.01	nc	<1	TVS	0
Total phosphorus, unfiltered (mg/L as P)	20	200	76	E 0.003	0.009	0.32	[4]0.05, 0.1	0
Orthophosphate, dissolved (mg/L as P)	10	112	55	<0.001	0.001	0.031	--	--
Yampa River and tributaries from Elk River confluence to Town of Hayden (Yampa River subwatershed 3; stream segments 2c, 11, 12, 13a, 13b, 13c, 13e, 13f, 20a)								
Nitrite, dissolved (mg/L as N)	8	122	62	<0.001	0.01	0.14	--	--
Nitrite, unfiltered (mg/L as N)	1	35	9	<0.0009	0.002	0.017	0.05	0
Nitrate, dissolved (mg/L as N)	29	373	76	<0.005	0.2	32	--	--
Nitrate, unfiltered (mg/L as N)	35	401	199	0.01	0.07	17	[3]10 (MCL), 100 (agriculture)	0
Total ammonia, unfiltered (mg/L as N)	25	356	213	<0.01	0.02	0.83	[3]TVS or none	0
Total phosphorus, unfiltered (mg/L as P)	37	418	41	<0.005	0.06	2.5	[4]0.05, 0.1	[5]3
Orthophosphate, dissolved (mg/L as P)	18	278	82	<0.001	0.01	0.2	--	--

Table 13. Summary statistics for selected nutrient water-quality data and Colorado Department of Public Health and Environment in-stream water-quality standards for stream sampling sites in the Upper Yampa River watershed and subwatersheds, Colorado, 1975 through 2009.—Continued

[No., number; mg/L, milligrams per liter; N, nitrogen; <, less than; --, no water-quality standard; MCL, maximum contaminant level; TVS, table value standard; P, phosphorus; E, estimated; nc, not computed. Number of significant figures for individual constituents may vary because data are from different sources and analytical periods. Water-quality standards are from Colorado Department of Public Health and Environment (2009a, 2010); standards are for aquatic-life protection, unless otherwise stated. See table 10 for sites in the Upper Yampa River watershed and subwatersheds with data for nutrients not in attainment of in-stream water-quality standards. Descriptions of stream segments are in Colorado Department of Public Health and Environment (2010). Collection of samples for nitrite analysis began in 1981]

Constituent (reporting units)	No. of sites	No. of samples	No. of censored values[1]	Minimum value[2]	Median value	Maximum value	In-stream water-quality standard	No. of sites with data not in attainment of standard
Yampa River and tributaries from Town of Hayden to Elkhead Creek confluence								
(Yampa River subwatershed 4; stream segments 2c, 12, 13d)								
Nitrite, dissolved (mg/L as N)	3	3	2	<0.001	<0.001	0.39	--	--
Nitrite, unfiltered (mg/L as N)	1	1	0	0.4	0.4	0.4	0.05	0
Nitrate, dissolved (mg/L as N)	7	83	6	<0.005	0.18	90	--	--
Nitrate, unfiltered (mg/L as N)	6	36	11	0.01	0.07	48	[3]10 (MCL), 100 (agriculture)	0
Total ammonia, unfiltered (mg/L as N)	6	31	19	<0.01	0.02	0.31	[3]TVS or none	0
Total phosphorus, unfiltered (mg/L as P)	11	56	3	<0.01	0.08	3.9	[4]0.05, 0.1	0
Orthophosphate, dissolved (mg/L as P)	4	66	24	<0.001	0.01	0.22	--	--
Elkhead Creek and tributaries								
(Elkhead Creek subwatershed; stream segments 14, 15, 20b)								
Nitrite, dissolved (mg/L as N)	4	179	146	E 0.003	nc	0.032	--	0
Nitrate, dissolved (mg/L as N)	7	195	83	<0.005	0.06	1.5	--	--
Nitrate, unfiltered (mg/L as N)	4	30	15	0.01	0.06	0.52	10 (MCL)	0
Total ammonia, unfiltered (mg/L as N)	2	20	16	<0.01	nc	0.15	TVS	0
Total phosphorus, unfiltered (mg/L as P)	6	198	44	0.004	0.02	0.923	[4]0.05, 0.1	[5]2
Orthophosphate, dissolved (mg/L as P)	6	184	136	E 0.009	<0.01	<0.18	--	--

[1]Censored values can be expressed as values less than the laboratory reporting level.

[2]For some constituents with censored data, the minimum censored value is greater than the minimum detected value that is shown.

[3]Water-quality standard varies by stream segment. See Colorado Department of Public Health and Environment (2010).

[4]Recommended concentration. See U.S. Environmental Protection Agency (2000). For stream segments with data for unfiltered total phosphorus, the recommended concentration has been applied to stream segments that have a standard for unfiltered nitrate.

[5]Number of sites with 85th percentile concentration greater than the U.S. Environmental Protection Agency recommended concentration.

[6]Seventeen unique site locations.

[7]Seven unique site locations.

for Yampa River at Milner (site 151). The median concentration was highest for April, which was likely a result of the initial flush of snowmelt that contains phosphorus bound to sediments (fig. 7C). This pattern is likely present for other streams in the UYRW.

Concentrations of unfiltered total phosphorus in about 14 percent (190 of 1,363) of individual samples were greater than USEPA recommended concentrations to control downstream eutrophication. This count excludes some stream segments in Yampa River subwatersheds 1, 3, and 4 because of naturally occurring high concentrations of phosphorus. A total of 59 samples from 7 sites (5 unique site locations) had unfiltered total phosphorus concentrations greater than 0.05 mg/L, the recommended concentration for streams that directly flow into lakes and reservoirs. Concentrations in 131 individual samples from 31 sites (29 unique site locations) were greater

than the recommended concentration of 0.1 mg/L for streams that do not directly flow into lakes and reservoirs. Fifty-one percent of concentrations greater than the recommendations were in samples collected during March, April, and May before and at the beginning of snowmelt runoff because phosphorus sorbs to particulate material. The 85th percentile concentration of total phosphorus data for 18 sites (17 unique site locations) exceeded the recommended concentrations of 0.05 or 0.1 mg/L; three of the streams flow directly into reservoirs (fig. 8, table 10). Data for eight of the sites were collected after 1994.

Unfiltered total phosphorus data for one site, Yampa River at Steamboat Springs (site 153), met the statistical requirements for trends testing. A statistically significant (p-value 0.04) trend in an upward direction was identified for unfiltered total phosphorus concentrations at the site for 1997 through 2008 (fig. 11, table 9). The rate of change was small, about 3.1 percent per year (magnitude about 0.001 mg/L per year). The median concentration of unfiltered total phosphorus at the site was 0.036 mg/L. The upward trend may reflect population growth and related land-use changes that have occurred upstream from the Steamboat Springs site.

Trace Elements and Uranium

For this study, trace elements are metallic and nonmetallic elements that generally occur in small (less than 1 mg/L) concentrations. Many trace elements are essential nutrients required by biota in small amounts, but substantial concentrations of trace elements can be toxic to aquatic life and possibly to wildlife, livestock, and people (Adriano, 2001). Some trace elements can bioaccumulate in biota and bioconcentrate in the food chain. Trace-element type and concentration in water are often directly related to natural sources such as soils, geology, geochemical conditions, and the presence of thermal springs. Streams in mineralized areas may contain high natural background concentrations of metals from the oxidation and weathering of minerals in rocks and soils. Common anthropogenic sources of trace elements in water are the deposition of metals released to the atmosphere from industrial activities and combustion, industrial water releases (particularly acidic mine drainage), and urban runoff.

Trace-element data for 2,427 samples collected from 1975 through 2009 at 145 sites in the UYRW are discussed in this report (tables 6, 7). Fifteen trace elements in the total recoverable form and (or) dissolved form are included (table 14). "Total recoverable" refers to that portion of a water and suspended-sediment sample measured by the total recoverable analysis procedure. Samples for trace-element

analysis were collected almost every year in Yampa River subwatersheds 1 and 3. The fewest trace-element samples were collected in the Elkhead Creek subwatershed (table 7).

More than two-thirds of the concentrations of dissolved and total recoverable cadmium, lead, nickel, and silver; dissolved chromium, copper, selenium, and zinc; and total dissolved and recoverable mercury were less than detection levels. Concentrations greater than 1,000 micrograms per liter (µg/L) were reported for total recoverable aluminum, iron, manganese, and zinc and dissolved iron, manganese, and strontium. Maximum concentrations of total recoverable aluminum, cadmium, and nickel; dissolved and total recoverable copper, iron, lead, manganese (fig. 6E), and mercury; and dissolved arsenic, boron, chromium, selenium, strontium, and zinc were detected in samples from Yampa River subwatersheds 3 and (or) 4. Maximum concentrations of total recoverable arsenic, chromium, strontium, and zinc were detected in samples from Yampa River subwatershed 1, and the maximum concentration of dissolved iron was in a sample from Yampa River subwatershed 2. These differences can likely be attributed to lithologic conditions in the subwatersheds. For example, iron can be associated with sedimentary and iron-rich igneous intrusive rocks (Colorado Department of Public Health and Environment, 2008). Data collected by the Seneca Coal Company indicate that the weathering and erosion of iron-containing lithologic formations and related soils contributes to elevated total recoverable iron concentrations in streams in the Grassy Creek area of Yampa River subwatershed 3. The mean content of manganese in the Wolf Creek coal bed in the eastern portion of the Yampa coal field was almost three times the mean content in other Cretaceous coal beds in the Yampa coal field (Brownfield and others, 1999). About 75 percent (89 of 118) of the samples analyzed

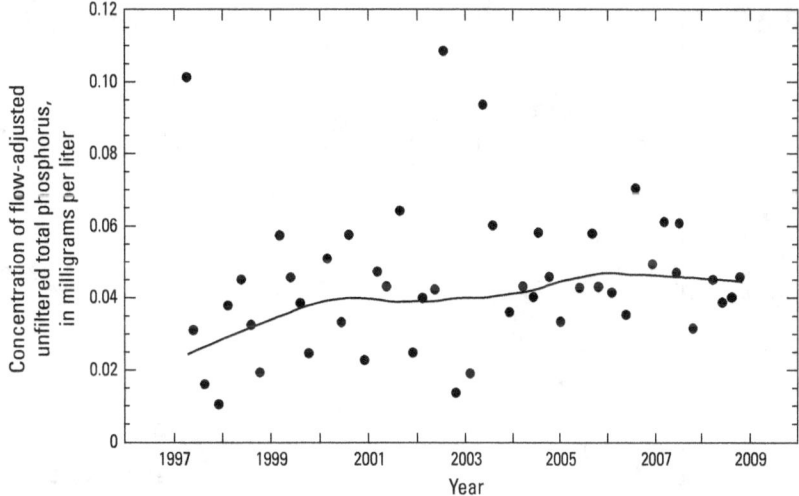

Figure 11. Concentrations of flow-adjusted unfiltered total phosphorus in Yampa River at Steamboat Springs (site 153), Upper Yampa River watershed, Colorado, 1997 through 2008.

Table 14.   Summary statistics for selected trace element and uranium water-quality data and Colorado Department of Public Health and Environment in-stream water-quality standards for stream sampling sites in the Upper Yampa River watershed and subwatersheds, Colorado, 1975 through 2009.

[No., number; --, no water-quality standard; <, less than; E, estimated; nc, not computed; TVS, table value standard; tr, trout; WS, water supply; SMCL, secondary maximum contaminant level; FRV, final residue value; sc, sculpin. All constituent values are reported in micrograms per liter. Number of significant figures for individual constituents may vary because data are from different sources and analytical periods. Water-quality standards are from Colorado Department of Public Health and Environment (2009a, 2010); standards are for aquatic-life protection, unless otherwise stated. Table value standards vary with stream hardness. See table 10 for sites in the Upper Yampa River watershed and subwatersheds with data for trace elements not in attainment of in-stream water-quality standards. Descriptions of stream segments are in Colorado Department of Public Health and Environment (2010)]

Constituent	No. of sites	No. of samples	No. of censored values[1]	Minimum value[2]	Median value	Maximum value	In-stream water-quality standard	No. of sites with data not in attainment of standard
					Upper Yampa River watershed			
Aluminum, dissolved	70	588	368	6	21	570	--	--
Aluminum, total recoverable	57	355	34	<1	310	83,000	--	--
Arsenic, dissolved	86	602	362	E 0.133	0.69	13	340 (acute)	0
Arsenic, total recoverable	25	31	17	<1	<1	3	[3]0.02, 7.6 (chronic)	0
Boron, dissolved	25	459	35	<10	60	630	750 (agriculture)	0
Cadmium, dissolved	113	1,006	957	E 0.013	nc	<3	[3]TVS(tr) (acute), TVS (acute, chronic, none	0
Cadmium, total recoverable	33	262	208	<0.02	0.02	9.2	--[4]	--[4]
Chromium, dissolved	36	154	144	<0.1	nc	50	[3]TVS (16 acute, 11 chronic) or none	0
Chromium, total recoverable	37	231	127	E 0.568	1.5	400	[3]100 (chronic) or none	0
Copper, dissolved	108	960	711	E 0.175	1.0	550	[3]TVS (acute, chronic) or none	4
Copper, total recoverable	102	977	337	E 0.403	2	510	[3]200 (acute) or none	0
Iron, dissolved	125	1,183	89	4	59	6,020	[3]300 WS (chronic, SMCL) or none	3 (WS)
Iron, total recoverable	123	2,029	13	<10	420	190,000	[3]1,000, 1,035, existing quality (chronic) or none	5
Lead, dissolved	108	925	883	0.04	nc	290	[3]TVS (acute, chronic) or none	0
Lead, total recoverable	96	545	432	E 0.03	0.46	300	[3]100 (chronic) or none	0
Manganese, dissolved	128	1,328	137	0.348	28	2,100	[3]TVS (acute, chronic), 50 WS (chronic, SMCL), none	[5]18 (WS)
Manganese, total recoverable	102	1,496	143	E 2.2	70	4,100	[3]200 (chronic) or none	0
Mercury, total, dissolved	67	397	379	0.008	nc	1.2	[3]0.01 FRV[6] (chronic) or none	0
Mercury, total recoverable	93	657	554	<0.0005	nc	2.1	[3]2.0 (acute) or none	0
Nickel, dissolved	45	135	108	<1	nc	<30	[3]TVS (acute, chronic) or none	0
Nickel, total recoverable	51	212	154	<1	2.4	280	[3]100, 200 (chronic) or none	[7]7
Selenium, dissolved	108	846	689	E 0.07	nc	300	[3]TVS (18.4 acute, 4.6 chronic) or none	0
Silver, dissolved	62	636	633	<0.008	nc	<3	[3]TVS (acute, chronic), TVS(tr) (chronic), none	0
Silver, total recoverable	39	172	166	<0.2	nc	<50	--[4]	--[4]
Strontium, dissolved	9	90	0	13	305	5,500	--	--
Strontium, total recoverable	3	28	0	30	300	400	--	--
Zinc, dissolved	122	1,036	708	E 0.546	1.8	960	[3]TVS (acute, chronic), TVS(sc) (chronic), none	0
Zinc, total recoverable	96	682	358	E 1.2	8.5	1,430	[3]2,000 (chronic) or none	0
Uranium, natural, dissolved	14	51	18	0.059	1	170	[8]30 (WS)	0

Table 14. Summary statistics for selected trace element and uranium water-quality data and Colorado Department of Public Health and Environment in-stream water-quality standards for stream sampling sites in the Upper Yampa River watershed and subwatersheds, Colorado, 1975 through 2009.—Continued

[No., number; --, no water-quality standard; <, less than; E, estimated; nc, not computed; TVS, table value standard; tr, trout; WS, water supply; SMCL, secondary maximum contaminant level; FRV, final residue value; sc, sculpin. All constituent values are reported in micrograms per liter. Number of significant figures for individual constituents may vary because data are from different sources and analytical periods. Water-quality standards are from Colorado Department of Public Health and Environment (2009a, 2010); standards are for aquatic-life protection, unless otherwise stated. Table value standards vary with stream hardness. See table 10 for sites in the Upper Yampa River watershed and subwatersheds with data for trace elements not in attainment of in-stream water-quality standards. Descriptions of stream segments are in Colorado Department of Public Health and Environment (2010)]

Constituent	No. of sites	No. of samples	No. of censored values[1]	Minimum value[2]	Median value	Maximum value	In-stream water-quality standard	No. of sites with data not in attainment of standard
							Yampa River and tributaries upstream from Chuck Lewis State Wildlife Area	
							(Yampa River subwatershed 1; stream segments 2a, 2c, 3, 4, 5, 6, 7)	
Aluminum, dissolved	19	196	156	6	nc	570	--	--
Aluminum, total recoverable	18	81	17	<1	105	2,500	--	--
Arsenic, dissolved	22	188	98	<1	<1	5	340 (acute)	0
Arsenic, total recoverable	14	19	8	<1	1	3	[3]0.02, 0.02–10, 7.6 (chronic)	0
Boron, dissolved	1	10	0	10	25	30	750 (agriculture)	0
Cadmium, dissolved	35	307	281	<0.02	nc	2	[3]TVS(tr) (acute), TVS (chronic), none	0
Cadmium, total recoverable	19	152	124	<0.3	nc	7.4	--[4]	--[4]
Chromium, dissolved	8	28	27	<1	nc	<20	[3]TVS (16 acute, 11 chronic) or none	0
Chromium, total recoverable	11	57	50	<1	nc	400	--[4]	--[4]
Copper, dissolved	33	330	258	<1	1.5	<20	[3]TVS (acute, chronic) or none	0
Copper, total recoverable	38	534	140	<0.6	1.9	32	--[4]	--[4]
Iron, dissolved	36	321	22	<10	71	1,008	[3]300 WS (chronic, SMCL) or none	2
Iron, total recoverable	41	915	2	10	480	8,970	[3]1,000 or none	1
Lead, dissolved	31	282	274	<0.08	nc	17	[3]TVS (acute, chronic) or none	0
Lead, total recoverable	34	202	161	<1	nc	<200	--[4]	--[4]
Manganese, dissolved	37	397	14	<1	37	450	[3]TVS (acute, chronic), 50 WS (chronic, SMCL), none	[9]10
Manganese, total recoverable	38	746	35	<2.8	76	1,217	--[4]	--[4]
Mercury, total, dissolved	15	64	61	<0.1	nc	<0.5	[3]0.01 FRV[6] (chronic) or none	0
Mercury, total recoverable	30	184	183	<0.0005	nc	<0.5	[3]2.0 (acute) or none	0
Nickel, dissolved	8	24	14	<1	<1	9	[3]TVS (acute, chronic) or none	0
Nickel, total recoverable	14	57	39	<1	2	<100	--[4]	--[4]
Selenium, dissolved	31	236	206	<0.2	nc	7.9	[3]TVS (18.4 acute, 4.6 chronic) or none	1
Silver, dissolved	20	208	207	<0.2	nc	2	[3]TVS (acute), TVS(tr) (chronic), none	0
Silver, total recoverable	21	67	65	<0.2	nc	<50	--[4]	--[4]
Strontium, dissolved	1	12	0	240	285	400	--	--
Strontium, total recoverable	2	20	0	280	315	400	--	--
Zinc, dissolved	31	290	197	<0.6	3.1	700	[3]TVS (acute, chronic), TVS(sc) (chronic), none	0
Zinc, total recoverable	35	201	138	<2	5.2	1,430	--[4]	--[4]
Uranium, natural, dissolved	4	21	7	0.88	1	170	[8]30 (WS)	0

Table 14. Summary statistics for selected trace element and uranium water-quality data and Colorado Department of Public Health and Environment in-stream water-quality standards for stream sampling sites in the Upper Yampa River watershed and subwatersheds, Colorado, 1975 through 2009.—Continued

[No., number; --, no water-quality standard; <, less than; E, estimated; nc, not computed; TVS, table value standard; tr, trout; WS, water supply; SMCL, secondary maximum contaminant level; FRV, final residue value; sc, sculpin. All constituent values are reported in micrograms per liter. Number of significant figures for individual constituents may vary because data are from different sources and analytical periods. Water-quality standards are from Colorado Department of Public Health and Environment (2009a, 2010); standards are for aquatic-life protection, unless otherwise stated. Table value standards vary with stream hardness. See table 10 for sites in the Upper Yampa River watershed and subwatersheds with data for trace elements not in attainment of in-stream water-quality standards. Descriptions of stream segments are in Colorado Department of Public Health and Environment (2010)]

Constituent	No. of sites	No. of samples	No. of censored values[1]	Minimum value[2]	Median value	Maximum value	In-stream water-quality standard	No. of sites with data not in attainment of standard
				Yampa River and tributaries from Chuck Lewis State Wildlife Area to Elk River confluence (Yampa River subwatershed 2; stream segments 2c, 3, 20a)				
Aluminum, dissolved	8	47	24	<30	34	170	--	--
Aluminum, total recoverable	6	10	0	56	505	2,600	340 (acute)	0
Arsenic, dissolved	13	67	57	<1	nc	2	0.02 (chronic)	0
Arsenic, total recoverable	4	5	4	<1	nc	1	--	0
Boron, dissolved	3	25	24	<10	nc	40	750 (agriculture)	0
Cadmium, dissolved	13	152	142	E 0.013	nc	<2	TVS(tr) (acute), TVS (chronic)	0
Cadmium, total recoverable	4	5	5	<0.3	<0.3	<0.3	--	--
Chromium, dissolved	11	45	43	<0.1	nc	<10	TVS (16 acute, 11 chronic)	0
Chromium, total recoverable	2	5	1	3	5.5	<10	--	--
Copper, dissolved	12	143	85	E 0.26	1.1	45.8	TVS (acute, chronic)	2
Copper, total recoverable	13	60	21	1.1	2.8	26.9	--	--
Iron, dissolved	21	106	1	17	80	6,020	300 WS (chronic, SMCL)	1
Iron, total recoverable	15	187	0	20	308	3,700	1,000 (chronic)	1
Lead, dissolved	12	139	119	E 0.04	nc	17	TVS (acute, chronic)	0
Lead, total recoverable	12	26	23	<1	nc	<200	--	--
Manganese, dissolved	22	184	23	1.3	16	133	TVS (acute, chronic), 50 WS (chronic, SMCL)	[10]5
Manganese, total recoverable	13	131	5	5	60	203	--	--
Mercury, total, dissolved	9	81	77	0.008	nc	<0.5	0.01 FRV[6] (chronic)	0
Mercury, total recoverable	9	43	39	<0.1	nc	<0.5	--	--
Nickel, dissolved	6	20	16	<1	<1	6	TVS (acute, chronic)	0
Nickel, total recoverable	5	16	15	3	nc	<50	--	--
Selenium, dissolved	13	146	122	E 0.07	nc	<6	TVS (18.4 acute, 4.6 chronic)	0
Silver, dissolved	9	119	118	<0.008	nc	1	TVS (acute), TVS(tr) (chronic)	0
Silver, total recoverable	4	5	5	<0.2	<0.2	<0.2	--	--
Strontium, dissolved	2	24	0	13	23.5	43	--	--
Zinc, dissolved	21	174	101	<0.6	2.1	587	[3]TVS (acute, chronic), TVS(sc) (chronic)	0
Zinc, total recoverable	10	25	13	<2	13.2	40	--	--
Uranium, natural, dissolved	1	4	3	<1	<1	1	[30] (WS)	0

Table 14. Summary statistics for selected trace element and uranium water-quality data and Colorado Department of Public Health and Environment in-stream water-quality standards for stream sampling sites in the Upper Yampa River watershed and subwatersheds, Colorado, 1975 through 2009.—Continued

[No., number; --, no water-quality standard; <, less than; E, estimated; nc, not computed; TVS, table value standard; tr, trout; WS, water supply; SMCL, secondary maximum contaminant level; FRV, final residue value; sc, sculpin. All constituent values are reported in micrograms per liter. Number of significant figures for individual constituents may vary because data are from different sources and analytical periods. Water-quality standards are from Colorado Department of Public Health and Environment (2009a, 2010); standards are for aquatic-life protection, unless otherwise stated. Table value standards vary with stream hardness. See table 10 for sites in the Upper Yampa River watershed and subwatersheds with data for trace elements not in attainment of in-stream water-quality standards. Descriptions of stream segments are in Colorado Department of Public Health and Environment (2010)]

Constituent	No. of sites	No. of samples	No. of censored values[1]	Minimum value[2]	Median value	Maximum value	In-stream water-quality standard	No. of sites with data not in attainment of standard
				Elk River and tributaries				
				(Elk River subwatershed; stream segments 8, 20a)				
Aluminum, dissolved	12	107	40	11.1	34	260	--	--
Aluminum, total recoverable	10	67	8	E 19.0	112	6,612	--	--
Arsenic, dissolved	13	83	64	E 0.133	nc	2	340 (acute)	0
Arsenic, total recoverable	2	2	2	<1	<1	<1	0.02 (chronic)	0
Boron, dissolved	1	1	1	<16	<16	<16	750 (agriculture)	0
Cadmium, dissolved	18	155	152	<0.02	nc	<2	TVS(tr) (acute), TVS (chronic)	0
Cadmium, total recoverable	3	21	18	<0.3	nc	2.9	--	--
Chromium, total recoverable	2	24	23	<1	nc	<20	--	--
Copper, dissolved	18	154	108	E 0.175	0.71	5.3	TVS (acute, chronic)	1
Copper, total recoverable	15	98	46	E 0.403	1	20	--	--
Iron, dissolved	17	127	1	E 6.4	54	420	300 WS (chronic, SMCL)	0
Iron, total recoverable	19	182	3	14	150	10,030	1000 (chronic)	0
Lead, dissolved	18	152	143	E 0.04	nc	13	TVS (acute, chronic)	0
Lead, total recoverable	15	98	65	E 0.03	0.31	<200	--	--
Manganese, dissolved	18	156	55	0.348	2.8	650	TVS (acute, chronic), 50 WS (chronic, SMCL)	0
Manganese, total recoverable	15	110	26	E 2.2	10	960	--	--
Mercury, total, dissolved	9	42	41	<0.1	nc	<0.5	0.01 FRV[6] (chronic)	0
Mercury, total recoverable	14	94	86	E 0.009	nc	<0.5	--	--
Nickel, dissolved	4	10	10	<1	nc	<2	TVS (acute, chronic)	0
Nickel, total recoverable	7	27	23	<1	nc	<100	--	--
Selenium, dissolved	12	88	86	<1	nc	<5	TVS (18.4 acute, 4.6 chronic)	0
Silver, dissolved	12	105	105	<0.2	nc	<1	TVS (acute), TVS(tr) (chronic)	0
Silver, total recoverable	4	17	17	<0.2	nc	<50	--	--
Strontium, dissolved	1	1	0	20.1	20.1	20.1	--	--
Strontium, total recoverable	1	8	0	30	65	110	--	--
Zinc, dissolved	18	158	120	E 0.546	1.05	52	[3]TVS (acute, chronic), TVS(sc) (chronic)	0
Zinc, total recoverable	15	98	66	E 1.20	3.6	100	--	--
Uranium, natural, dissolved	4	12	4	0.059	0.499	6	[8]30 (WS)	0

Table 14. Summary statistics for selected trace element and uranium water-quality data and Colorado Department of Public Health and Environment in-stream water-quality standards for stream sampling sites in the Upper Yampa River watershed and subwatersheds, Colorado, 1975 through 2009.—Continued

[No., number; --, no water-quality standard; <, less than; E, estimated; nc, not computed; TVS, table value standard; tr, trout; WS, water supply; SMCL, secondary maximum contaminant level; FRV, final residue value; sc, sculpin. All constituent values are reported in micrograms per liter. Number of significant figures for individual constituents may vary because data are from different sources and analytical periods. Water-quality standards are from Colorado Department of Public Health and Environment (2009a, 2010); standards are for aquatic-life protection, unless otherwise stated. Table value standards vary with stream hardness. See table 10 for sites in the Upper Yampa River watershed and subwatersheds with data for trace elements not in attainment of in-stream water-quality standards. Descriptions of stream segments are in Colorado Department of Public Health and Environment (2010)]

Constituent	No. of sites	No. of samples	No. of censored values[1]	Minimum value[2]	Median value	Maximum value	In-stream water-quality standard	No. of sites with data not in attainment of standard
							Yampa River and tributaries from Elk River confluence to Town of Hayden (Yampa River subwatershed 3; stream segments 2c, 11, 12, 13a, 13b, 13c, 13e, 13f, 20a)	
Aluminum, dissolved	23	172	111	8	18	<250	--	--
Aluminum, total recoverable	15	119	9	<1	1,500	68,000		--
Arsenic, dissolved	26	162	78	0.4	1	13	340 (acute)	0
Arsenic, total recoverable	4	4	2	<1	<1	2	[3]0.02, 7.6	0
Boron, dissolved	12	345	8	<10	60	630	750 (agriculture)	0
Cadmium, dissolved	33	265	260	<0.02	nc	<3	[3]TVS(tr) (acute), TVS (acute, chronic), none	0
Cadmium, total recoverable	4	81	58	<0.02	0.03	9.2	--[4]	--[4]
Chromium, dissolved	14	44	39	<0.8	nc	50	[3]TVS (16 acute, 11 chronic), none	0
Chromium, total recoverable	14	80	30	<0.8	6	54	[3]100 (chronic) or none	0
Copper, dissolved	32	218	207	<1	nc	550	[3]TVS (acute, chronic) or none	1
Copper, total recoverable	23	182	110	<0.6	3.0	510	[3]200 (acute) or none	0
Iron, dissolved	37	446	45	4	50	2,300	300 WS (chronic, SMCL) or none	0
Iron, total recoverable	32	522	8	<10	460	190,000	[3]1,000, 1,035, existing quality (chronic) or none	3
Lead, dissolved	34	237	234	<0.08	nc	<30	[3]TVS (acute, chronic) or none	0
Lead, total recoverable	22	158	139	<1	nc	300	[3]100 (chronic) or none	0
Manganese, dissolved	37	394	29	<4	50	1,500	[3]TVS (acute, chronic), 50 WS (chronic, SMCL), none	3
Manganese, total recoverable	23	331	71	<2.8	120	4,100	[3]200 (chronic) or none	0
Mercury, total, dissolved	24	130	126	<0.01	nc	1.2	[3]0.01 FRV[6] (chronic) or none	0
Mercury, total recoverable	26	214	154	<0.0005	0.06	2.1	--	--
Nickel, dissolved	19	51	51	<1	nc	<30	[3]TVS (acute, chronic) or none	0
Nickel, total recoverable	18	62	62	<1	nc	<100	[3]200 (chronic) or none	0
Selenium, dissolved	38	247	178	<0.4	<1	290	[3]TVS (18.4 acute, 4.6 chronic) or none	[1]14
Silver, dissolved	14	119	118	<0.06	nc	<3	[3]TVS (acute, chronic), TVS(tr) (chronic), none	0
Silver, total recoverable	7	46	42	<0.2	nc	<50	[3]TVS (acute, chronic), TVS(sc) (chronic), none	--
Strontium, dissolved	3	38	0	290	1,020	5,500	--	--
Zinc, dissolved	38	289	208	<0.6	1.1	280	[3]2,000 (chronic) or none	0
Zinc, total recoverable	22	261	103	<2	20	1,000	[8]30 (WS)	0
Uranium, natural, dissolved	4	13	4	<1	1	4		0

Table 14. Summary statistics for selected trace element and uranium water-quality data and Colorado Department of Public Health and Environment in-stream water-quality standards for stream sampling sites in the Upper Yampa River watershed and subwatersheds, Colorado, 1975 through 2009.—Continued

[No, number; --, no water-quality standard; <, less than; E, estimated; nc, not computed; TVS, table value standard; tr, trout; WS, water supply; SMCL, secondary maximum contaminant level; FRV, final residue value; sc, sculpin. All constituent values are reported in micrograms per liter. Number of significant figures for individual constituents may vary because data are from different sources and analytical periods. Water-quality standards are from Colorado Department of Public Health and Environment (2009a, 2010); standards are for aquatic-life protection, unless otherwise stated. Table value standards vary with stream hardness. See table 10 for sites in the Upper Yampa River watershed and subwatersheds with data for trace elements not in attainment of in-stream water-quality standards. Descriptions of stream segments are in Colorado Department of Public Health and Environment (2010)]

Constituent	No. of sites	No. of samples	No. of censored values[1]	Minimum value[2]	Median value	Maximum value	In-stream water-quality standard	No. of sites with data not in attainment of standard
Yampa River and tributaries from Town of Hayden to Elkhead Creek confluence (Yampa River subwatershed 4; stream segments 2c, 12, 13d)								
Aluminum, dissolved	6	47	27	8	12	<250	--	--
Aluminum, total recoverable	5	41	0	130	2,200	83,000	--	--
Arsenic, dissolved	6	43	18	<1	1	4	340 (acute)	0
Boron, dissolved	4	66	0	50	80	590	750 (agriculture)	0
Cadmium, dissolved	8	47	43	<0.02	nc	<2	[3]TVS(tr) (acute),	0
Chromium, dissolved	1	1	1	<0.8	<0.8	<0.8	TVS (acute, chronic), none	0
Chromium, total recoverable	6	29	3	<0.8	4	30	[3]TVS (16 acute, 11 chronic), none	0
Copper, dissolved	7	42	27	1.1	1.9	60	[3]100 (chronic) or none	0
Copper, total recoverable	5	55	6	<0.6	5.9	250	[3]TVS (acute, chronic) or none	0
Iron, dissolved	8	107	14	<10	40	1,718	[3]200 (acute) or none	0
Iron, total recoverable	10	145	0	50	690	190,000	[3]300 WS (chronic, SMCL) or [3]1,000, existing quality (chronic) or none	0
Lead, dissolved	7	38	37	<0.08	nc	290	[3]TVS (acute, chronic) or none	0
Lead, total recoverable	5	12	11	<1	nc	<200	[3]100 (chronic) or none	0
Manganese, dissolved	8	109	2	5	30	2,100	[3]TVS (acute, chronic), 50 WS (chronic, SMCL), none	0
Manganese, total recoverable	7	119	0	20	80	3,300	[3]200 (chronic) or none	0
Mercury, total, dissolved	5	22	21	<0.01	nc	<0.5	[3]0.01 FRV[6] (chronic) or none	0
Mercury, total recoverable	8	64	40	<0.0005	0.08	<0.5	--	--
Nickel, dissolved	3	10	10	<1	nc	<2	[3]TVS (acute, chronic) or none	0
Nickel, total recoverable	2	6	5	<50	nc	280	[3]200 (chronic) or none	0
Selenium, dissolved	7	47	23	<0.4	1	300	[3]TVS (18.4 acute, 4.6 chronic) or none	2
Silver, dissolved	3	26	26	<0.2	nc	<0.5	[3]TVS (acute, chronic), TVS(tr) (chronic), none	0
Zinc, dissolved	7	43	18	<0.6	10	960	[3]TVS (acute, chronic), TVS(sc), none	0
Zinc, total recoverable	6	50	6	<2	40	890	[3]2,000 (chronic) or none	0
Uranium, natural, dissolved	1	1	0	4	4	4	[8]30 (WS)	0
Elkhead Creek and tributaries (Elkhead Creek subwatershed; stream segments 14, 15, 20b)								
Aluminum, dissolved	2	19	10	<15	46	190	--	--
Aluminum, total recoverable	3	37	0	50.5	390	9,063	--	--
Arsenic, dissolved	6	59	47	<1	nc	<5	340 (acute)	0
Arsenic, total recoverable	1	1	1	<1	<1	<1	0.02 (chronic)	0
Boron, dissolved	3	12	2	<10	20	60	750 (agriculture)	0
Cadmium, dissolved	6	80	79	<0.04	nc	<2	TVS(tr) (acute), TVS (chronic)	0

Table 14. Summary statistics for selected trace element and uranium water-quality data and Colorado Department of Public Health and Environment in-stream water-quality standards for stream sampling sites in the Upper Yampa River watershed and subwatersheds, Colorado, 1975 through 2009.—Continued

[No., number; --, no water-quality standard; <, less than; E, estimated; nc, not computed; TVS, table value standard; tr, trout; WS, water supply; SMCL, secondary maximum contaminant level; FRV, final residue value; sc, sculpin. All constituent values are reported in micrograms per liter. Number of significant figures for individual constituents may vary because data are from different sources and analytical periods. Water-quality standards are from Colorado Department of Public Health and Environment (2009a, 2010); standards are for aquatic-life protection, unless otherwise stated. Table value standards vary with stream hardness. See table 10 for sites in the Upper Yampa River watershed and subwatersheds with data for trace elements not in attainment of in-stream water-quality standards. Descriptions of stream segments are in Colorado Department of Public Health and Environment (2010)]

Constituent	No. of sites	No. of samples	No. of censored values[1]	Minimum value[2]	Median value	Maximum value	In-stream water-quality standard	No. of sites with data not in attainment of standard
				Elkhead Creek and tributaries				
			(Elkhead Creek subwatershed; stream segments 14, 15, 20b)—Continued					
Cadmium, total recoverable	3	3	3	<0.3	<0.3	<0.3	--	--
Chromium, dissolved	2	36	34	E 0.507	nc	5.5	TVS (16 acute, 11 chronic)	0
Chromium, total recoverable	2	36	20	E 0.568	1.32	10.0	--	--
Copper, dissolved	6	73	26	E 0.856	1.7	<5	TVS (acute, chronic)	0
Copper, total recoverable	8	48	14	<1	2.23	23.1	--	--
Iron, dissolved	6	76	6	6	30.5	620	300 WS (chronic, SMCL)	0
Iron, total recoverable	6	78	0	30	428	17,690	1,000 (chronic)	0
Lead, dissolved	6	77	76	E 0.048	nc	<5	TVS (acute, chronic)	0
Lead, total recoverable	8	49	33	E 0.539	1.09	<200	--	--
Manganese, dissolved	6	88	14	<1	10	100	TVS (acute, chronic), 50 WS (chronic, SMCL)	0
Manganese, total recoverable	6	59	6	8.46	37	523	--	--
Mercury, total, dissolved	5	58	53	<0.01	nc	<0.5	0.01 FRV[6] (chronic)	0
Mercury, total recoverable	6	58	52	<0.011	nc	<0.5	--	--
Nickel, dissolved	5	20	7	<1	2	3	TVS (acute, chronic)	0
Nickel, total recoverable	5	44	10	E 1.31	2.85	<50	--	--
Selenium, dissolved	7	82	74	<0.4	nc	<5	TWS (18.4 acute, 4.6 chronic)	0
Silver, dissolved	4	59	59	<0.1	nc	<2	TVS (acute), TVS(tr) (chronic)	0
Silver, total recoverable	3	37	37	<0.2	nc	<1	--	--
Strontium, dissolved	2	15	0	130	270	503	--	--
Zinc, dissolved	7	82	64	<0.6	2.1	28	TVS (acute, chronic)	0
Zinc, total recoverable	8	47	32	<2	9.5	68.8	--	--

[1]Censored values can be expressed as values less than the laboratory reporting level.

[2]For some constituents with censored data, the minimum censored value is greater than the minimum detected value that is shown.

[3]Water-quality standard varies by stream segment. See Colorado Department of Public Health and Environment (2010).

[4]No data were available for stream segment(s) listed in appendix 2 that have a Colorado Department of Public Health and Environment water-quality standard for the constituent of interest.

[5]Sixteen unique site locations.

[6]Final Residue Value (FRV) is the maximum allowed concentration of total mercury in water that will present bioaccumulation or bioconcentration of methylmercury in edible fish tissue (Colorado Department of Public Health and Environment, 2009a).

[7]Six unique site locations.

[8]Uranium level in surface water used for water supply shall be maintained at the lowest practical level. The maximum allowed concentration in water used for supply is 30 micrograms per liter, unless naturally occurring concentrations are greater (Colorado Department of Public Health and Environment, 2009a).

[9]Nine unique site locations.

[10]Four unique site locations.

[11]Three unique site locations.

for dissolved and total recoverable strontium had concentrations greater than the 60 µg/L, which is the median strontium concentration for major rivers in North America (Hem, 1992). Concentrations greater than 60 µg/L were present in samples from every subwatershed except Yampa River subwatershed 4, and no strontium data were available for this subwatershed (table 14). The Wadge and Wolf Creek coal beds in the Yampa coal field have high contents of strontium (30 times greater than the average value for Cretaceous-age coal) (Affolter, 2000).

Seasonal variation in trace-element concentrations was evident for total recoverable concentrations of aluminum, copper, iron, and zinc for Yampa River at Milner (site 151) (fig. 7D), which is likely representative of other sites. The highest 5 to 10 percent of the concentration values of all four trace elements were detected in April and May. Concentrations were elevated during the initial pulse of snowmelt runoff, and to a lesser extent later in the snowmelt period, as a result of particulate-phase trace elements binding to sediment as material is washed off the land surface. The seasonal variation in total recoverable manganese concentrations was less evident. The seasonal pattern in dissolved iron concentrations was similar to that of total recoverable iron. For other dissolved trace elements, a seasonal pattern in concentrations was not visually discernible. Sufficient data for trend testing were available only for total recoverable iron and dissolved and total recoverable manganese for Yampa River at Steamboat Springs (site 153). No statistically significant trends (p-values 0.53, 0.17) were identified for total recoverable iron or dissolved manganese for data collected from 1997 through 2008. A statistically significant downward trend (p-value 0.03) was identified for concentrations of total recoverable manganese in samples collected from 1997 through 2008 (table 9). The rate of change was small, 4.6 percent per year (magnitude of 2.6 µg/L per year). The median concentration of total recoverable manganese at the site was 45.6 µg/L.

Concentrations of dissolved cadmium, manganese, selenium, and silver; dissolved and total recoverable arsenic, chromium, copper, lead, nickel, and zinc; and total recoverable iron and mercury were compared to CDPHE standards for the protection of aquatic life (table 14, appendix 2). Dissolved boron concentrations were compared to the CDPHE agricultural-use standard. Concentrations of total dissolved mercury were compared to the final residue value, which is the maximum allowed concentration of total mercury in water that will present bioaccumulation or bioconcentration of methylmercury in edible fish tissue (Colorado Department of Public Health and Environment, 2009a). Although concentrations of some trace elements in individual samples were greater than the CDPHE standards, about 90 percent (130 of 145) of stream sites with trace element data were in attainment of standards for dissolved boron, cadmium, manganese, and silver; dissolved and total recoverable arsenic, chromium, lead, mercury,

nickel, and zinc; and total recoverable copper because the 50th (total recoverable) and 85th (dissolved) percentiles of the concentration data for a trace element at a particular site were less than the respective standard.

A total of 15 sites (14 unique site locations) with data for dissolved copper, total recoverable iron, and (or) dissolved selenium were not in attainment of the respective CDPHE aquatic-life standards (fig. 12, table 10). All data for 11 of 15 sites were collected before 1999. Attainment of the standard for dissolved copper was not met for four sites (fig. 12, table 10). The two sites in the Yampa River subwatershed 2 and the one site in the Elk River subwatershed naturally have low hardness; mean hardness for the three sites was 15 mg/L or less. Because aquatic-life standards for many trace elements are hardness dependent, extremely low hardness for a sample results in a smaller value for the standard compared to a sample with harder water. The CDPHE aquatic-life standard for total recoverable iron was not met for five sites, all data at these sites were collected before 1999 (fig. 12, table 10). Nonattainment of the standard prior to 1999 may not be a current issue of concern; attainment of the standard was met for all sites that had total recoverable iron data collected after 1998. Individual samples with total recoverable iron concentration greater than the standard were most commonly collected from March through May when increased sediment load would be carried in streams with the flush of snowmelt runoff. The CDPHE chronic aquatic-life standard for dissolved selenium of 4.6 µg/L was not met for seven sites (six unique site locations); the acute standard of 18.4 µg/L was not met for six of the seven sites (fig. 12, table 10). All sites but one are in Yampa River subwatersheds 3 and 4, portions of which are underlain by seleniferous Lewis Shale.

Attainment of the CDPHE water-supply standards for dissolved iron and (or) dissolved manganese were not met at a total of 19 sites overall (17 unique site locations). Three sites in Yampa River subwatersheds 1 or 2 were not in attainment of the CDPHE water-supply standard for dissolved iron, and 18 sites (16 unique sites) in Yampa River subwatershed 1, 2, or 3 were not in attainment of the water-supply standard for dissolved manganese (fig. 12, table 10). Elevated concentrations of dissolved iron and manganese could be the result of lithologic conditions or possible reduction-oxidation processes that may occur in groundwater. Iron is the fourth most abundant element in the Earth's crust (Lutgens and others, 2012). It is associated with a variety of minerals in igneous and sedimentary rocks and is present in organic materials. Manganese commonly is in the same minerals as iron (Driver and others, 1984). As a result, large amounts of iron and manganese can be common in surface water and groundwater.

Two nonreservoir stream segments in the UYRW are on the state of Colorado 2012 303(d) list of impaired waters for nonattainment of aquatic-life standards because of trace element contamination (Colorado Department of Public Health

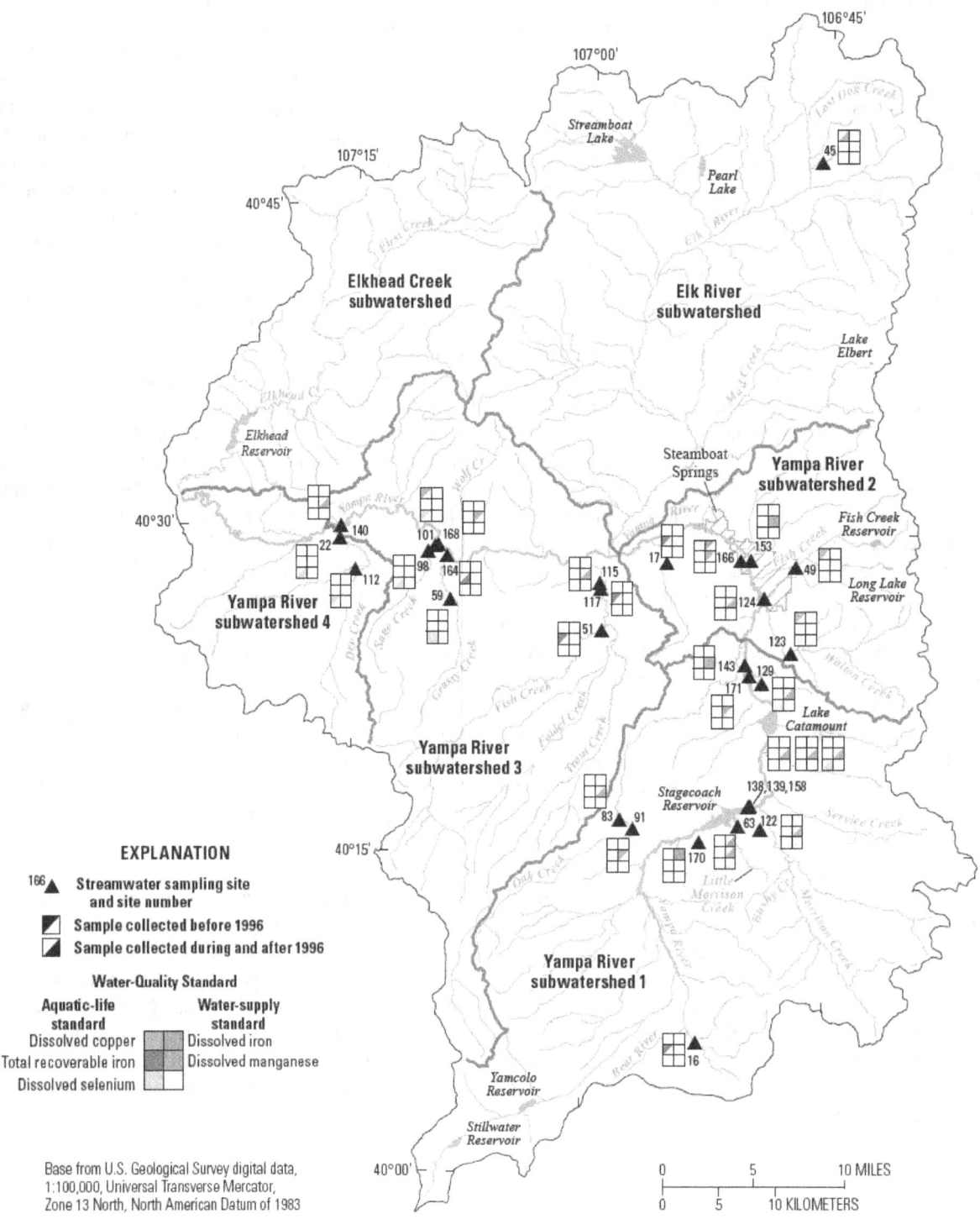

Figure 12. Location of water-quality sampling sites with exceedances of Colorado Department of Public Health and Environment in-stream water-quality standards for trace elements, Upper Yampa River watershed, Colorado, 1975 through 2009.

and Environment, 2012). All or a portion of segment 13d for Dry Creek is listed for total recoverable iron (snowmelt season) or selenium, and a portion of Sage Creek in segment 13e is listed for selenium (table 5) . Elkhead Creek is provisionally listed for aquatic life (table 5) (Colorado Department of Public Health and Environment, 2012). Portions of four stream segments are on the state of Colorado monitoring and evaluation list for dissolved iron and lead, manganese, mercury, selenium, or zinc contamination (table 5) (Colorado Department of Public Health and Environment, 2012).

Uranium, a radiochemical, is a human carcinogen that also can be harmful to kidneys (U.S. Environmental Protection Agency, 2012a). Water-quality samples collected from streams in the UYRW were infrequently analyzed for uranium. Dissolved uranium data were available for only 51 samples collected at 14 sites, most commonly in Yampa River subwatersheds 1 and 3 and the Elk River subwatershed (table 7). Detected concentrations of dissolved uranium ranged from 0.059 to 170 µg/L (table 14). About 35 percent of the uranium data were censored at detection levels of 1 or 5 µg/L. Attainment of the CDPHE water-supply standard of 30 µg/L was met for all sites with uranium data; the 85th percentile of the concentration data for each of the 12 sites with uranium data was less than 30 µg/L.

Uranium is an important source of radon gas, another human carcinogen. The potential for radon in indoor air increases with higher concentrations of uranium in rocks (U.S. Environmental Protection Agency, 2012b). The USEPA has assigned Garfield, Moffat, and Rio Blanco Counties to the radon-potential category of 1, signifying a high potential for elevated indoor radon levels (U.S. Environmental Protection Agency, 2010a). Routt County is in the radon-potential category of 2 and has a moderate potential for elevated indoor radon levels. In the UYRW database, no water-quality data for radon were available for analysis.

Coliform Bacteria

Coliforms are bacteria present in the digestive tracts of warm-blooded animals and in soil and vegetation. The bacteria are not likely to cause illness themselves, but their presence in water indicates that disease-causing pathogens could also be in the water. Total coliform consists of a large group of different types of bacteria. Fecal coliforms are bacteria that are present in the feces and intestines of warm-blooded animals. Their presence in water indicates recent contamination by animal waste or sewage. *Escherichia coli* (*E. coli)* are a subgroup of fecal coliform bacteria and are the only coliform group with a CDPHE water-quality standard for recreational use of water. The numeric value of the standard varies by stream segment. Potential sources of fecal coliform in the UYRW include, but are not limited to, recreational water users, wildlife, livestock, and septic systems. Concentrations of fecal coliforms are typically affected by, but not limited to, temperature, salinity, light intensity, rainfall, and streamflow (Chigbu and Sobeler, 2007).

In the UYRW, data for coliform bacteria (total, fecal, and *E. coli*) were available for 432 samples collected from 89 stream sites (table 6) during 1975 and 1976 and 1989 through 2009 (table 7). Samples were collected most frequently and from the greatest number of stream sites in Yampa River subwatershed 2. The fewest samples and sites were in Yampa River subwatershed 4.

Total coliform data were available for 119 samples collected during 1975 and 1976 at 43 sites, primarily in Yampa River subwatershed 2. Concentrations ranged from no detection to 18,000 colonies per 100 milliliters (col/100 mL) (table 15); about 80 percent (95 of 119) of the concentrations were less than 400 col/100 mL. Most (20 of 24) concentrations of 400 col/100 mL or more were detected in samples from Yampa River subwatershed 2. Fecal coliform data were available for 399 samples collected from 88 sites in multiple subwatersheds (table 15). Collection dates were during 1975,1976, and most years from 1989 through 2008. About 95 percent (378 of 399) of the fecal coliform concentrations were less than 200 col/100 mL. Water-quality standards have not been established for total and fecal coliform.

A total of 122 samples collected at 6 sites from 1991 through 2009 were analyzed for *E. coli*. Concentrations ranged from less than 1 to 733 col/100 mL, and the median was 18 col/100 mL (table 15). Recent (2000 through 2009) *E. coli* data are from samples collected at Elk River near Milner, Colo. (site 33), Fish Creek at Upper Station (site 49),Yampa River at Steamboat Springs (site 153), and two Elkhead Creek sites (sites 40, 42). The 2010 CDPHE recreation standard for *E. coli* of 126 col/100 mL was exceeded in five samples collected during 1994, 1999, 2001, and 2003. Two samples were from Yampa River at Steamboat Springs (site 153), and three samples were from two Elkhead Creek sites (sites 40, 42). The Yampa River in the vicinity of Steamboat Springs is heavily used for recreation during summer, and the Elkhead Creek subwatershed has a high percentage of rangeland and pasture. No sites with an *E. coli* standard of 205 or 630 col/100 mL were sampled. Attainment of the *E. coli* standard is based on the geometric mean of representative stream samples. For this study, an insufficient number of *E. coli* samples were collected to calculate geometric means for comparison to the standard of 126 col/100 mL. Comparison of individual *E. coli* concentrations to the standard, however, can give a general indication of water quality in streams where the bacteria are present. Portions of three stream segments in the UYRW are on the CDPHE 303(d) list of impaired waters or monitoring and evaluation list for *E. coli* (table 5) (Colorado Department of Public Health and Environment, 2012).

Suspended Sediment

Suspended sediment or suspended solids are very fine particles suspended in water for a substantial time period without settling. This includes silt and soil from erosion and storm and urban runoff; remains from the breakdown of terrestrial

Table 15. Summary statistics for selected coliform bacteria water-quality data and Colorado Department of Public Health and Environment in-stream water-quality standards for stream sampling sites in the Upper Yampa River watershed and subwatersheds, Colorado, 1975, 1976, and 1989 through 2009.

[No , number; --, no standard; <, less than. All coliform values are reported in colonies per 100 milliliters. Water-quality standards are from Colorado Department of Public Health and Environment (2009a, 2010). Descriptions of stream segments are in Colorado Department of Public Health and Environment (2010)]

Coliform bacteria	No. of sites	No. of samples	No. of censored values[1]	Minimum value	Median value	Maximum value	In-stream water-quality standard	No. of samples with data exceeding standard
Upper Yampa River watershed								
Total coliform	43	119	12	0	60	18,000	--	--
Fecal coliform	88	399	52	0	14	2,100	--	--
Escherichia coli	6	122	5	<1	18	733	[2]126/100	5
Yampa River and tributaries upstream from Chuck Lewis State Wildlife Area (Yampa River subwatershed 1; stream segments 2a, 2c, 3, 4, 5, 6, 7)								
Total coliform	9	29	2	<1	20	750	--	--
Fecal coliform	20	102	29	<1	4	560	--	--
Yampa River and tributaries from Chuck Lewis State Wildlife Area to Elk River confluence (Yampa River subwatershed 2; stream segments 2c, 3, 20a)								
Total coliform	19	59	4	<1	92	18,000	--	--
Fecal coliform	32	150	11	0	15	1,800	--	--
Escherichia coli	2	64	6	<1	12	177	126/100	2
Elk River and tributaries (Elk River subwatershed; stream segments 8, 20a)								
Total coliform	5	16	5	<1	4	1,700	--	--
Fecal coliform	8	34	7	<1	2	1,200	--	--
Escherichia coli	1	6	0	7	22	108	126/100	0
Yampa River and tributaries from Elk River confluence to Town of Hayden (Yampa River subwatershed 3; stream segments 2c, 11, 12, 13a, 13b, 13c, 13e, 13f, 20a)								
Total coliform	7	7	0	0	96	200	--	--
Fecal coliform	18	32	2	0	13	133	--	--
Yampa River and tributaries from Town of Hayden to Elkhead Creek confluence (Yampa River subwatershed 4; stream segments 2c, 12, 13d)								
Total coliform	1	1	0	210	210	210	--	--
Fecal coliform	4	12	0	2	14	261	--	--
Elkhead Creek and tributaries (Elkhead Creek subwatershed; stream segments 14, 15, 20b)								
Total coliform	2	7	1	<1	24	190	--	--
Fecal coliform	6	69	3	<1	29	2,100	--	--
Escherichia coli	3	52	0	1	21	733	126/100	3

[1]Censored values can be expressed as values less than the laboratory reporting level.

[2]No *Escherichia coli* data were available for stream segments with standards of 205 or 630 colonies per 100 milliliters.

and aquatic biota; and wastes from industry and water-treatment plants (Murphy, 2007b). The amount and size of suspended sediment in water are affected by streamflow. More and larger suspended material can be carried in water with a higher streamflow than with a lower streamflow (fig. 13). Because phosphorus adsorbs to the surface of sediment, phosphorus concentrations are higher when suspended-sediment concentrations are higher (fig. 13). In the UYRW, a total of 1,079 suspended-sediment samples were collected from 65 sites during most years from 1975 through 2003 (tables 6, 7). Sample collection occurred after 1992 only in the Elk River and Elkhead Creek subwatersheds. Suspended-sediment concentrations typically were lowest from August through February when streamflow was lowest. Concentrations typically were higher during May in the Elk River and Elkhead Creek subwatersheds and during April in other subwatersheds. The highest concentrations (greater than 2,000 mg/L) were in samples collected from streams in areas with sedimentary rocks, especially Yampa River subwatersheds 3 and 4 (table 16).

Lakes and Reservoirs

The UYRW water-quality database contains data for 42 lake and reservoir sites in 30 bodies of water for 369 sample days for the period 1975 through 2009 (table 3). Sites with different names and identifiers in the database but with the same latitude and longitude are counted as one site. Samples

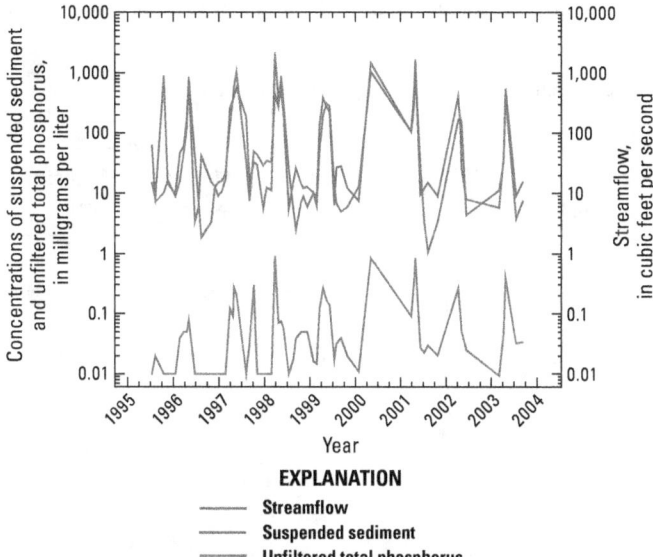

Figure 13. Streamflow and concentrations of suspended sediment and unfiltered total phosphorus in Elkhead Creek above Long Gulch (site 40), Upper Yampa River watershed, Colorado, July 1995 through September 2003.

collected at multiple depths on the same day are counted as a one sample day. For 26 of the lake and reservoir sites in 16 bodies of water, data were collected for a maximum of 4 sample days, all before 1986; about one-half of these sites were described as potholes or ponds. Data also were available for 16 sites and 327 samples collected on 180 sample days from 5 lakes and reservoirs for the period 1983 through 2009 (appendix 4). The analysis of lake and reservoir data was restricted to these five water bodies because data collection was long term (greater than 10 years) and (or) occurred after 1994. Lake Elbert and Long Lake Reservoir (fig. 1), high-altitude lakes in and near the Mount Zirkel Wilderness Area north and east of Steamboat Springs, are long-term data-collection sites. Samples analyzed for this study were collected from Lake Elbert on 74 sample days from 1985 through 2009 and from Long Lake Reservoir on 64 sample days from 1985 through 2005. Samples were collected in Stagecoach Reservoir on 43 sample days at 5 sites during 1990 through 1992 and July 2006. Data for Steamboat Lake were available for only one sample day in July 2006. Samples were collected in Elkhead Reservoir on 18 days at 8 sites from July 1995 through August 2001.

Lake and reservoir sites were grouped for analysis on the basis of site and data similarities. Data for Lake Elbert and Long Lake Reservoir were analyzed together. Analysis of data for Stagecoach Reservoir was limited to samples collected on July 25, 2006, at one site. These data were compared to data collected in Steamboat Lake on the same day. Water quality of Elkhead Reservoir was analyzed in detail by Kuhn and others (2003) and is summarized later in this subsection.

Lake Elbert and Long Lake Reservoir were sampled two-to-five times per year during the open-water season, and grab samples were collected manually from the shore at locations with little or no vegetation (Mast and others, 2005). One sample per day for each reservoir was analyzed for this study. Both lakes were very dilute. Median specific conductance values for near-surface samples were 10.6 µS/cm for Lake Elbert and 18.3 µS/cm for Long Lake Reservoir (table 17). All values were 34 µS/cm or less, which reflects the igneous-rock composition in the drainage basins of the water bodies. Specific conductance in both water bodies typically was lower during snowmelt runoff when lake water was diluted than during summer when groundwater discharge had a greater effect on water chemistry. Median values of pH near the water surface were similar (7.1 and 7.2) for the two water bodies. On three samples days for Long Lake Reservoir, values of pH in three near-surface water samples were less than the CDPHE water-quality minimum standard of 6.5 for aquatic-life protection. Two samples from Long Lake Reservoir had dissolved oxygen concentrations equal to 0.2 mg/L. Hardness, ANC, and sulfate concentrations were slightly higher in Long Lake Reservoir than in Lake Elbert, reflecting differences in biogeochemistry of the two water bodies and the quality of the shallow groundwater system underlying the water bodies (Mast and others, 2005). Low ANC (17.9 mg/L or less) indicates that both water bodies are sensitive to acidic deposition. A statistically significant decrease in sulfate concentrations in Lake Elbert has been observed for the period 1985 through 2008 (Mast and others, 2011). This decrease has been driven by reductions in sulfur dioxide emissions from a nearby upwind power plant and a related decrease in atmospheric deposition of sulfate (Mast and others, 2011). Most (123 of 136) nitrate concentrations in near-surface samples from both water bodies were less than detection levels. Mast and others (2005) report that the Lake Elbert and Long Lake Reservoir watersheds are less sensitive to atmospheric deposition of nitrogen than reservoirs in the Front Range of Colorado because of a greater capacity for nitrogen assimilation. Unfiltered total phosphorus concentrations in water samples from both water bodies were 0.02 mg/L or less. Water-supply standards for dissolved manganese and dissolved iron were exceeded on one and two sample days, respectively, for Long Lake Reservoir.

Depth-profile measurements for July 26, 2006, for one site near the dam in Stagecoach Reservoir and one site in Steamboat Lake were compiled for specific conductance, pH, water temperature, and dissolved oxygen (figs. 14A, 14B). Samples also were collected near the water surface and near the bottom of the water column for analysis of other physical properties, major ions, nutrients, trace elements, and chlorophyll a. Values for selected physical properties and constituents in the near surface and near bottom samples are shown in table 18.

Table 16.   Summary statistics for suspended-sediment water-quality data for stream sampling sites in the Upper Yampa River watershed and subwatersheds, Colorado, 1975 through 2003.

[No., number. All suspended-sediment values are reported in milligrams per liter. Number of significant figures for suspended-sediment data may vary because data are from different sources and analytical periods.]

Watershed or subwatershed	No. of sites	No. of samples	No. of censored values[1]	Minimum value	Median value	Maximum value
Upper Yampa River watershed	65	1,079	11	0	40	11,300
Subwatershed						
Yampa River and tributaries upstream from Chuck Lewis State Wildlife Area (Yampa River subwatershed 1)	17	261	0	0	33	2,340
Yampa River and tributaries from Chuck Lewis State Wildlife Area to Elk River confluence (Yampa River subwatershed 2)	14	66	0	0	7	676
Elk River and tributaries (Elk River subwatershed)	10	101	11	0	4	565
Yampa River and tributaries from Elk River confluence to Town of Hayden (Yampa River subwatershed 3)	12	399	0	0	85	11,300
Yampa River and tributaries from Town of Hayden to Elkhead Creek confluence (Yampa River subwatershed 4)	5	92	0	8.3	90	10,200
Elkhead Creek and tributaries (Elkhead Creek subwatershed)	7	160	0	1	16	2,197

[1]Censored values can be expressed as values less than the laboratory reporting level.

Table 17.   Summary statistics for selected water-quality data and Colorado Department of Public Health and Environment in-stream water-quality standards for sampling sites in Lake Elbert and Long Lake Reservoir, Upper Yampa River watershed, Colorado.

[No., number; µS/cm; microsiemens per centimeter at 25 degrees Celsius; --, no water-quality standard; mg/L, milligrams per liter; nd, not determined; $CaCO_3$, calcium carbonate; WS, water supply; N, nitrogen; <, less than; nc, not computed; P, phosphorus; µg/L, micrograms per liter; TVS, table value standard. Samples from Lake Elbert were collected from 1985 through 2009. Samples from Long Lake Reservoir were collected from 1985 through 2005. Summary statistics are calculated from samples collected near the water surface. Water-quality standards are from Colorado Department of Public Health and Environment (2009a, 2010); standards are for aquatic-life protection, unless otherwise stated. Descriptions of stream segments are in Colorado Department of Public Health and Environment (2010)]

Physical property or constituent (reporting units)	No. of sample days	No. of censored values[1]	Minimum value	Median value	Maximum value	In-stream water-quality standard	No. of sample days with data exceeding standard
Lake Elbert (stream segment 1b)							
Specific conductance (µS/cm)	34	0	7.2	10.6	13.3	--	--
pH (standard units)	28	0	6.7	7.1	7.6	6.5–9.0	0
Dissolved oxygen (mg/L)	25	0	5.9	6.5	7.5	6.0	nd[2]
Hardness (mg/L as $CaCO_3$)	71	0	2.6	3.4	5.1	--	--
Acid neutralizing capacity (mg/L as $CaCO_3$)	29	0	2.6	3.5	4.4	--	--
Sulfate, dissolved (mg/L)	73	0	0.307	0.504	0.700	250 (WS)	0
Nitrate, dissolved (mg/L as N)	73	70	<0.006	nc	0.014	10	0
Total phosphorus, unfiltered (mg/L as P)	35	5	[3]0.001	<0.005	0.02	--	--
Iron, dissolved (µg/L)	31	1	<3	20	98	300 (WS)	0
Iron, total recoverable (µg/L)	30	0	20	85	370	1,000	0
Manganese, dissolved (µg/L)	32	12	<1	1	4	TVS, 50 (WS)	0
Long Lake Reservoir (stream segment 2b)							
Specific conductance, laboratory (µS/cm)	38	0	14.0	18.3	34.0	--	--
pH (standard units)	32	0	6.2	7.2	8.2	6.5–9.0	3
Dissolved oxygen (mg/L)	10	0	0.2	6.4	7.4	6.0	nd[2]
Hardness (mg/L as $CaCO_3$)	63	0	5.4	7.5	15.8	--	--
Acid neutralizing capacity (mg/L as $CaCO_3$)	39	0	2.1	6.1	17.9	--	--
Sulfate, dissolved (mg/L)	63	0	0.892	1.3	2.5	250 (WS)	0
Nitrate, dissolved (mg/L as N)	63	53	<0.007	nc	0.11	10	0
Total phosphorus, unfiltered (mg/L as P)	2	0	0.017	0.018	0.02	--	--
Iron, dissolved (µg/L)	42	0	44	140	400	300 (WS)	2 (WS)
Manganese, dissolved (µg/L)	42	3	<1	4	150	TVS, 50 (WS)	1 (WS)

[1]Censored values can be expressed as values less than the laboratory reporting level.

[2]The number of exceedances of the water-quality standard could not be determined. Application of the dissolved oxygen standard is dependent on depth of water. Sample depth was not recorded for most samples with dissolved oxygen data.

[3]The minimum censored value is greater than the minimum detected value.

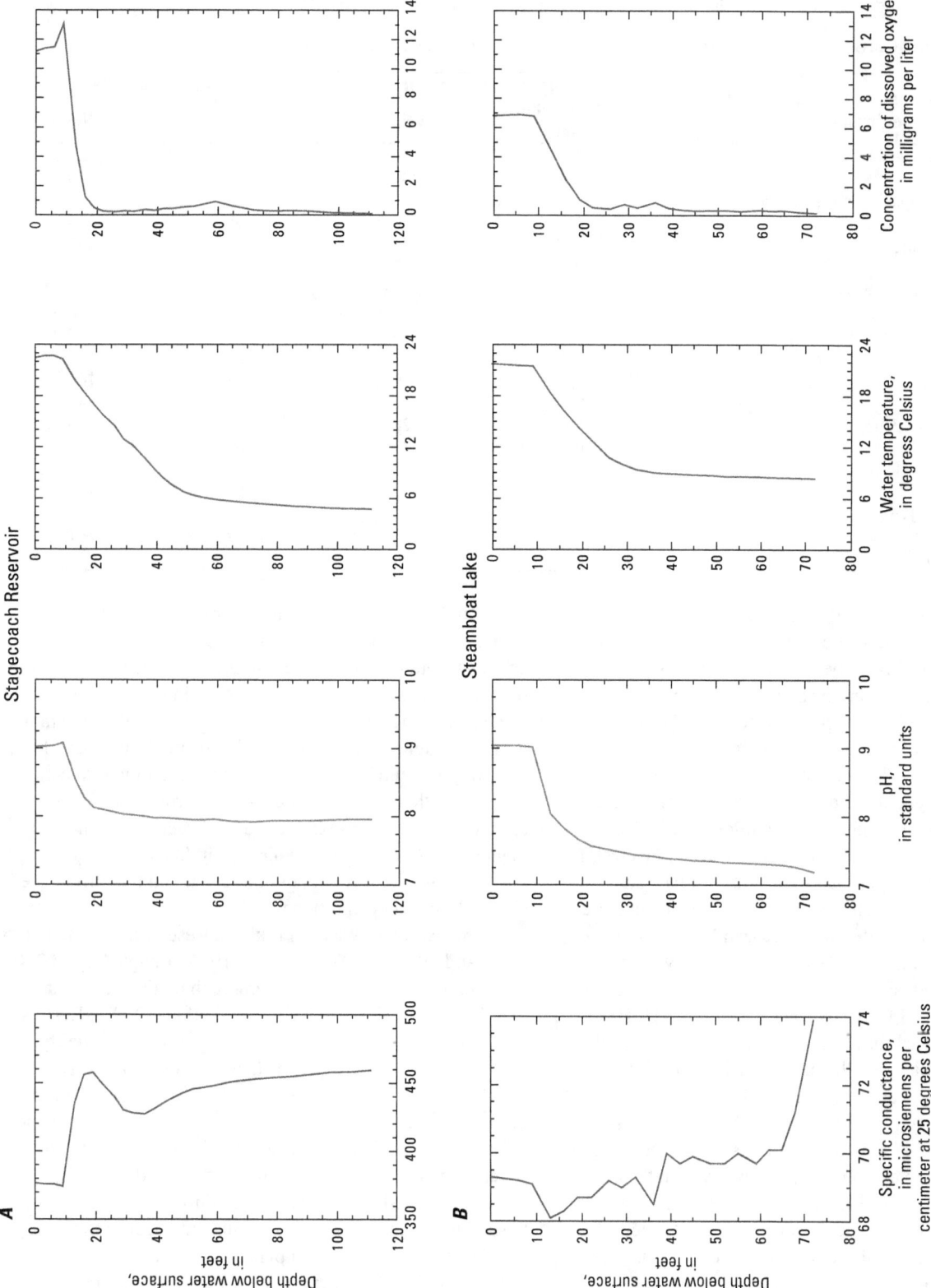

Figure 14. Depth profiles of specific conductance, pH, water temperature, and dissolved oxygen concentrations for (A) Stagecoach Reservoir and (B) Steamboat Lake, Upper Yampa River watershed, Colorado, July 25, 2006.

Table 18. Concentrations of selected physical properties and constituents in near-surface and near-bottom water-quality samples, Stagecoach Reservoir and Steamboat Lake, Upper Yampa River watershed, Colorado, July 25, 2006.

[mg/L, milligrams per liter; $CaCO_3$, calcium carbonate; <, less than; N, nitrogen; P, phosphorus; µg/L micrograms per liter]

Physical property or constituent (reporting units)	Stagecoach Reservoir		Steamboat Lake	
	Sample depth (feet)	Value	Sample depth (feet)	Value
Hardness (mg/L as $CaCO_3$)	0	170	3	29
	104	200	68	30
Acid neutralizing capacity (mg/L as $CaCO_3$)	0	150	3	34
	104	180	68	35
Sulfate, unfiltered (mg/L)	0	46	3	<3
	104	62	68	4
Nitrate, unfiltered (mg/L as N)	0	<0.02	3	<0.02
	104	0.35	68	0.08
Total phosphorus, unfiltered (mg/L as P)	0	0.04	3	0.02
	104	0.14	68	0.09
Iron, dissolved (µg/L)	0	<10	3	140
	104	17	68	540
Iron, total recoverable (µg/L)	0	28	3	230
	104	37	68	810
Manganese, dissolved (µg/L)	0	<2	3	2
	104	160	68	110
Selenium, dissolved (µg/L)	0	<1	3	<1
	104	1.1	68	<1
Chlorophyll a (µg/L)	0	8,200	3	800

Vertical stratification within the water column was evident for specific conductance, pH, water temperature, and dissolved oxygen (fig. 14). Specific conductance was much higher for Stagecoach Reservoir than for Steamboat Lake. Most of the watershed for Stagecoach Reservoir overlies sedimentary rocks, which contribute dissolved materials to surface water (fig. 5). The watershed for Steamboat Lake primarily overlies igneous and metamorphic rocks that are more resistant to weathering and dissolution (fig. 5). Water clarity, as measured by Secchi depth transparency (Wetzel, 1983), was lower in Stagecoach Reservoir (4.4 ft) than in Steamboat Lake (7.9 ft). This is probably due to an increased amount of particulate matter in Stagecoach Reservoir as compared to Steamboat Lake. Water column pH was lower (about 7.2–7.5) in Steamboat Lake than in Stagecoach Reservoir (7.9–9.1) (fig. 14). No exceedances of the CDPHE water-quality standard for pH of 6.5–9.0 were observed. Water temperature at the site near the dam in each reservoir was similar at the surface (about 22 °C) and lower (less than 6 °C) at depth in Stagecoach Reservoir than in Steamboat Lake (8.4 °C or greater) (fig. 14). Anoxic conditions (dissolved oxygen concentrations less than 0.5 mg/L) at depth were indicated for both bodies of water (fig. 14). Loss of oxygen primarily is due to oxygen consumption at the sediment-water interface where bacterial decomposition of sediment organic matter is greatest and to the use of oxygen by aquatic organisms in the water column (Wetzel, 1983).

Concentrations of chemical constituents in both bodies of water were lower near the water surface than near the bottom of the water column (table 18). Concentrations of unfiltered nitrate less than detection levels in near-surface samples

probably indicate biological uptake of the nutrient. Hardness, ANC, unfiltered sulfate, and total phosphorus in near-surface and near-bottom samples were higher in Stagecoach Reservoir, probably because of sedimentary rocks in the reservoir drainage basin. Concentrations of dissolved and total recoverable iron were higher in Steamboat Lake than in Stagecoach Reservoir, which could be attributed to the hydrologic interactions of water with igneous and metamorphic rocks in the reservoir drainage basin. Chlorophyll a was an order of magnitude higher in Stagecoach Reservoir than in Steamboat Lake. With only one sample, it is not possible to explain this difference in chlorophyll a concentration.

The quality of water in Elkhead Reservoir was studied by Kuhn and others (2003) from July 1995 through August 2001. Information presented in this paragraph on the water quality of the reservoir is from Kuhn and others (2003). Results from the study indicate seasonal changes in water quality for many physical properties and chemical constituents. Depth-profile measurements showed that the reservoir was stratified during summer and late winter and mixed during spring and fall. Minimum values of specific conductance (138 to 169 µS/cm) occurred during snowmelt inflow, and maximum values (424 to 610 µS/cm) occurred during early spring prior to snowmelt runoff. Values of pH indicate neutral to slightly alkaline conditions. Median pH near the water surface ranged from 7.2 to 8.0. Water temperature was lowest (about 0 °C) during winter and warmest (about 20 °C) during summer. During stratification, median dissolved oxygen concentrations were about 7.1 and 7.2 for samples from near the water surface and 4.8 to 5.6 mg/L for near-bottom samples. Some concentrations in near-bottom samples were less than 0.5

mg/L and indicate anoxic conditions. Water transparency was greatest after snowmelt because of the settling of suspended sediment. Concentrations of most nutrients in near-surface and near-bottom samples were highest during snowmelt inflow. Total phosphorus concentrations were highest at depth during July and September. Median concentrations of chlorophyll *a* were 1.1 µg/L or less at all sites. The trophic state of the reservoir ranged from oligotrophic (nutrient poor) to eutrophic (nutrient rich). Concentrations of nitrogen and phosphorous in 52 percent of samples indicate that phosphorus was the limiting nutrient; concentrations in 9 percent of samples indicate that nitrogen was the limiting nutrient. Fish consumption advisories for mercury have been established for Elkhead Reservoir and Lake Catamount, south of Steamboat Springs (*http://www.cdphe.state.co.us/wq/FishCon/*, accessed June 2012) (table 5).

Groundwater

Water-quality data for groundwater in the UYRW were available for 328 wells (table 3). Six wells were sampled by the CDOA during 1998, and 322 wells were sampled by the USGS from 1975 through 1989 (fig. 15, table 3, appendix 5). The sampled wells are concentrated in the middle latitudes of the watershed. About 39 percent (128 of 328) of the wells were located in the Yampa coal field (fig. 15). A total of 1,580 samples were collected—6 by the CDOA and 1,574 by the USGS. For the USGS wells, 810 samples had water-quality data and 764 samples had data only for water-level measurements. About 59 percent (479 of 816) of the water-quality samples were collected from wells located in the Yampa coal field. Each CDOA well and about 66 percent (212 of 322) of the USGS wells with water-quality data were sampled once. Samples with water-quality data most often were collected during 1975, 1978, and 1988 (table 19). All water-quality samples for groundwater, with the exception of one, had data for physical properties. The fewest data were available for organic carbon, stable isotopes (carbon, hydrogen, oxygen, sulfur), and radiochemical constituents (potassium-40, tritium) (table 19). Analysis of groundwater-quality data for this study focuses on one sample per day for physical properties, dissolved solids, major ions, nutrients, and trace elements. Multiple water-quality samples collected on the same day at a site are not included in the analysis.

For all CDOA wells and about 30 percent (97 of 322) of the USGS wells, no aquifer or geologic unit information was provided with the well-construction information. In this report, these wells are grouped in the category "unknown geologic units" (table 20). Wells with a geologic unit description tap aquifers in 12 geologic units, most commonly the flood-plain alluvium and Mesaverde Group (table 20). Water-quality samples most often were collected from aquifers in the unknown geologic units, Mesaverde Group, and terrace alluvium. Most (88 percent or more) of the samples collected from the unknown geologic units and the Mesaverde Group were from wells located in the Yampa coal field (table 20). The oldest geologic

units with water-quality data presented in this report are the Precambrian erathem and the Jurassic Curtis Formation of the San Rafael Group. In ascending order, the stratigraphy of the Upper Cretaceous- and Tertiary-age formations with water-quality data are the Mancos Shale, Mesaverde Group (Iles and William Fork Formations), Lewis Shale, Fort Union Formation, and Browns Park Formation. The Upper Cretaceous and Eocene series represents rocks from particular time periods, but no information on individual geologic units is available from the aquifer descriptions. The Mancos Shale predominately is mudrock that formed in the marine environment of the Western Interior Seaway (Hettinger and Kirschbaum, 2002). It grades into the Mesaverde Group which consists of marine and nonmarine deposits of sandstone, shale, and coal beds that formed with the regression and transgression of the Western Interior Seaway (Brownfield and others, 1999). The Lewis Shale is composed of shale, siltstone, and smaller amounts of sandstone that formed in an offshore marine environment. The Tertiary Fort Union Formation is composed of sandstones, conglomerates, shales, and coal (Colson, 1969). The Browns Park Formation includes riverine sandstone, conglomerates, and siltstone that eroded from nearby mountains and volcanic ash (Covay and Tobin, 1981). Flood-plain and terrace alluvium and valley fill are Quaternary unconsolidated deposits of gravel, sand, silt, and clay along streams and in valleys and are derived from rock and soil erosion.

The quality of groundwater is a function of various physical and geochemical processes, including precipitation, the depositional environment and type of aquifer sediments that groundwater moves through, groundwater age, dissolution of soluble minerals in rocks and soils, ion exchange reactions, and reduction-oxidation processes. Reports with information on groundwater quality in the UYRW include Brodgen and Giles (1977), Covay and Tobin (1981), Robson and Stewart (1990), and Topper and others (2003). Additional reports are listed in Driver and others (1984).

Physical Properties

Data for physical properties of groundwater were available for 815 water-quality samples collected from 328 wells (table 19). Water temperature and specific conductance data were the most common, and dissolved oxygen concentrations were the least common (table 21). Specific conductance ranged from 50 to 15,900 µS/cm with a median of 1,170 µS/cm (table 21). Median specific conductance was lowest (163 µS/cm) in samples from the Precambrian erathem, a geologic unit with igneous and metamorphic rocks (fig. 5, table 21). Specific conductance also typically was lower in samples from the Browns Park Formation than from the Lewis and Mancos Shales and Mesaverde Group (fig. 16), probably reflecting the riverine depositional environment of the Browns Park Formation and the marine or marine-nonmarine depositional environment of the latter three geologic units. Values of pH in UYRW groundwater samples ranged from 5.3 to 12.3, and the median was 7.6. About 14 percent (94 of 676) of pH

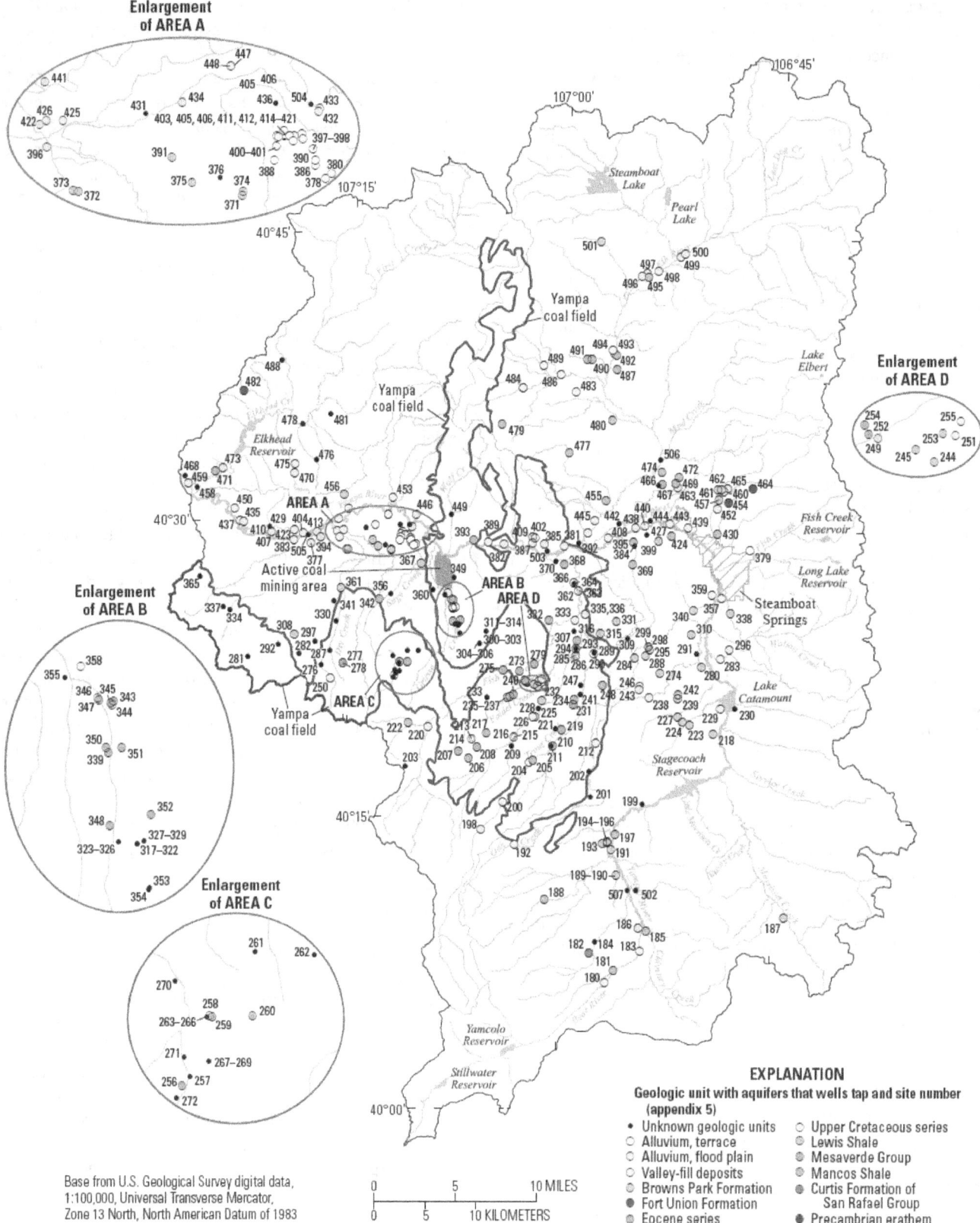

Figure 15. Location of groundwater wells completed in selected geologic units, Upper Yampa River watershed, Colorado, 1975 through 1989 and 1998.

Table 19. Period of water-quality record and number of water-quality samples collected per year from groundwater wells, by physical properties or constituent group, Upper Yampa River watershed Colorado, 1975 through 1989 and 1998.

[--, no data]

Physical properties or constituent group	Number of water-quality samples collected																
	1975	1976	1977	1978	1979	1980	1981	1982	1983	1984	1985	1986	1987	1988	1989	1998	Total
Upper Yampa River watershed																	
Physical properties	145	10	19	146	91	45	57	47	29	2	--	3	22	[1]121	72	6	815
Dissolved solids	56	5	18	45	39	42	57	47	29	2	--	3	22	[1]121	72	6	564
Major ions	56	5	18	45	61	43	57	47	29	2	--	3	22	[1]121	72	6	587
Nutrients	56	5	18	45	35	43	55	47	29	--	--	3	22	[1]120	10	6	494
Trace elements	56	5	18	46	84	43	57	47	29	2	--	3	22	[1]121	72	4	609
Radiochemical	--	--	--	--	--	9	35	18	--	--	--	--	--	22	4	--	88
Organic carbon	--	--	--	--	1	--	1	--	--	--	--	--	--	--	4	--	6
Stable isotopes	--	--	--	--	--	--	--	--	--	--	--	--	--	23	4	--	27
Total number of samples	145	10	19	147	91	45	57	47	29	2	--	3	22	[1]121	72	6	

[1]Count does not include multiple (14) samples collected on one day from selected wells.

Table 20. Number of wells sampled and groundwater-quality samples collected from selected geologic units in the Upper Yampa River watershed, Colorado, 1975 through 1989 and 1998.

[No., number; --, no data]

Geologic unit near well screen	Geologic time	Primary lithology	No. of wells	No. of groundwater-quality samples[1]
Alluvium, flood plain	Quaternary	Unconsolidated material	75	85
Alluvium, terrace	Quaternary	Unconsolidated material	23	122
Browns Park Formation	Tertiary	Sandstone	29	29
Curtis Formation of San Rafael Group	Jurassic	Sandstone, shale	1	1
Eocene series	Tertiary	Shale, sandstone	1	2
Fort Union Formation	Tertiary	Sandstone, shale, coal	1	1
Lewis Shale	Cretaceous	Shale, sandstone	14	19
Mancos Shale	Cretaceous	Shale, sandstone	23	26
Mesaverde Group	Cretaceous	Sandstone, shale, coal	46	142
Precambrian erathem	Precambrian	Crystalline rocks	2	2
Upper Cretaceous series	Cretaceous	Sandstone, shale, coal	3	3
Valley-fill deposits	Quaternary	Unconsolidated material	7	10
Unknown	--	--	103	374
Total number of wells and groundwater-quality samples			328	816

[1]Count does not include multiple (14) water-quality samples collected on one day from selected wells.

values did not meet the CDPHE SMCL (table 21) (Colorado Department of Public Health and Environment, 2009b). Values less than the standard minima of 6.5 were not as common as values greater than the standard maxima of 8.5 (table 21). Values less than the standard minima were measured in 14 samples from 13 wells, most commonly in 14 percent of samples from the flood-plain alluvium and 14 percent of wells completed in the valley-fill deposits. Values of pH greater than standard maxima were measured in 80 samples from 38 wells, most commonly in 16 percent of samples from unknown geologic units and 29 percent of wells from the Mesaverde Group. Less than 4 percent of wells in the Yampa coal field had sample pH less than 6.5. About 80 percent of all wells with sample pH greater than 8.5 were in the Yampa coal field. Water temperature ranged from 2 to 50 °C with a median of

10.5 °C (table 21). Dissolved oxygen data were available for 116 samples from 33 wells, and concentrations ranged from 0 to 10 mg/L (table 21). High values of ANC in some samples from the sedimentary rock units indicate that the sample water was well buffered.

Reduction-oxidation (redox) processes can be an extremely valuable tool in interpreting water-quality data and assessing the susceptibility or vulnerability of an aquifer to contamination. Redox processes, including dissolved oxygen reduction and nitrate, iron, manganese, and sulfate reduction, affect water chemistry and water quality in all groundwater systems. Groundwater with dissolved oxygen concentrations greater than or equal to 0.5 mg/L (oxic conditions) can be more susceptible to nitrate contamination than other groundwater. Groundwater with dissolved oxygen concentrations

Table 21. Summary statistics for selected physical property water-quality data and exceedances of Colorado Department of Public Health and Environment water-quality standards for groundwater in the Upper Yampa River watershed, Colorado, 1975 through 1989 and 1998

[No., number; Min., minimum; Max., maximum; μS/cm, microsiemens per centimeter at 25 degrees Celsius; --, no water-quality standard; SMCL, secondary maximum contaminant level; C, degrees Celsius; mg/L, milligrams per liter; CaCO₃, calcium carbonate; ne, no exceedance. Number of significant figures for individual constituents may vary among the geologic units because data are from different sources and different analytical periods. Water-quality standards are from Colorado Department of Public Health and Environment (2009b)]

Physical property (reporting units)	Total no. of wells	Total no. of samples	Total no. of censored values	Min. value	Median value	Max. value	Water-quality standard	Exceedances of water-quality standard			
								No. of wells	No. of samples	Year(s)	Range of values not meeting standard
Upper Yampa River watershed											
Specific conductance (μS/cm)	264	715	0	50	1,170	15,900	--	--	--	--	--
pH (standard units)	257	676	0	5.3	7.6	12.3	6.5 minima (SMCL)	13	14	[1]1975-84	5.3-6.4
							8.5 maxima (SMCL)	38	80	[1]1975-89	8.6-12.3
Water temperature (°C)	327	731	0	2	10.5	50	--	--	--	--	--
Dissolved oxygen (mg/L)	33	116	0	0	1.4	10	--	--	--	--	--
Hardness (mg/L as CaCO₃)	201	574	0	4.9	330	7,000	--	--	--	--	--
Acid neutralizing capacity (mg/L as CaCO₃)	165	264	0	19	260	2,760	--	--	--	--	--
Alluvium, flood plain											
Specific conductance (μS/cm)	43	45	0	77	370	1,850	--	--	--	--	--
pH (standard units)	40	42	0	5.8	7.2	9.4	6.5 minima (SMCL)	5	6	1978	5.8-6.4
							8.5 maxima (SMCL)	6	6	1978	8.7-9.4
Water temperature (°C)	75	84	0	4	11	30	--	--	--	--	--
Dissolved oxygen (mg/L)	1	1	0	4.6	4.6	4.6	--	--	--	--	--
Hardness (mg/L as CaCO₃)	15	16	0	19	170	550	--	--	--	--	--
Acid neutralizing capacity (mg/L as CaCO₃)	14	15	0	19	191	522	--	--	--	--	--
Alluvium, terrace											
Specific conductance (μS/cm)	23	121	0	50	974	2,700	--	--	--	--	--
pH (standard units)	23	88	0	7.0	7.6	8.5	6.5 minima (SMCL)	0	0	ne	ne
							8.5 maxima (SMCL)	0	0	ne	ne
Water temperature (°C)	23	84	0	3	11	22	--	--	--	--	--
Hardness (mg/L as CaCO₃)	22	43	0	140	440	1,100	--	--	--	--	--
Acid neutralizing capacity (mg/L as CaCO₃)	22	87	0	110	180	580	--	--	--	--	--
Browns Park Formation											
Specific conductance (μS/cm)	28	28	0	50	455	1,475	--	--	--	--	--
pH (standard units)	27	27	0	6.4	7.5	9.2	6.5 minima (SMCL)	1	1	1978	6.4
							8.5 maxima (SMCL)	1	1	1978	9.2
Water temperature (°C)	28	28	0	5	9	22	--	--	--	--	--
Hardness (mg/L as CaCO₃)	14	14	0	10	180	428	--	--	--	--	--
Acid neutralizing capacity (mg/L as CaCO₃)	7	7	0	150	210	700	--	--	--	--	--
Curtis Formation of San Rafael Group											
Specific conductance (μS/cm)	1	1	0	430	430	430	--	--	--	--	--
pH (standard units)	1	1	0	7.3	7.3	7.3	6.5 minima (SMCL)	0	0	ne	ne
							8.5 maxima (SMCL)	0	0	ne	ne
Water temperature (°C)	1	1	0	38	38	38	--	--	--	--	--
Hardness (mg/L as CaCO₃)	1	1	0	180	180	180	--	--	--	--	--
Acid neutralizing capacity (mg/L as CaCO₃)	1	1	0	210	210	210	--	--	--	--	--

Table 21. Summary statistics for selected physical property water-quality data and exceedances of Colorado Department of Public Health and Environment water-quality standards for groundwater in the Upper Yampa River watershed, Colorado, 1975 through 1989 and 1998.—Continued

[No., number; Min., minimum; Max., maximum; µS/cm, microsiemens per centimeter at 25 degrees Celsius; °C, degrees Celsius; mg/L, milligrams per liter; CaCO3, calcium carbonate; ne, no exceedance. Number of significant figures for individual constituents may vary among the geologic units because data are from different sources and different analytical periods. Water-quality standards are from Colorado Department of Public Health and Environment (2009b)]

Physical property (reporting units)	Total no. of wells	Total no. of samples	Total no. of censored values	Min. value	Median value	Max. value	Water-quality standard	Exceedances of water-quality standard: No. of wells	No. of samples	Year(s)	Range of values not meeting standard
Eocene series											
Specific conductance (µS/cm)	1	1	0	800	800	800	--	--	--	--	--
pH (standard units)	1	1	0	8.3	8.3	8.3	6.5 minima (SMCL)	0	0	ne	ne
							8.5 maxima (SMCL)	0	0	ne	ne
Water temperature (°C)	1	2	0	16	16	16	--	--	--	--	--
Hardness (mg/L as CaCO3)	1	1	0	12	12	12	--	--	--	--	--
Acid neutralizing capacity (mg/L as CaCO3)	1	1	0	366	366	366	--	--	--	--	--
Fort Union Formation											
Specific conductance (µS/cm)	1	1	0	1,125	1,125	1,125	--	--	--	--	--
pH (standard units)	1	1	0	6.5	6.5	6.5	6.5 minima (SMCL)	0	0	ne	ne
							8.5 maxima (SMCL)	0	0	ne	ne
Water temperature (°C)	1	1	0	14	14	14	--	--	--	--	--
Hardness (mg/L as CaCO3)	1	1	0	390	390	390	--	--	--	--	--
Acid neutralizing capacity (mg/L as CaCO3)	1	1	0	75	75	75	--	--	--	--	--
Lewis Shale											
Specific conductance (µS/cm)	9	11	0	470	1,200	4,000	--	--	--	--	--
pH (standard units)	9	11	0	6.9	8.0	9.0	6.5 minima (SMCL)	0	0	ne	ne
							8.5 maxima (SMCL)	1	1	1978	9.0
Water temperature (°C)	14	19	0	9	12	21	--	--	--	--	--
Hardness (mg/L as CaCO3)	6	6	0	24	370	630	--	--	--	--	--
Acid neutralizing capacity (mg/L as CaCO3)	6	7	0	189	255	1,190	--	--	--	--	--
Mancos Shale											
Specific conductance (µS/cm)	23	25	0	75	766	3,800	--	--	--	--	--
pH (standard units)	23	24	0	6.1	7.4	8.7	6.5 minima (SMCL)	1	1	1978	6.1
							8.5 maxima (SMCL)	1	1	1975	8.7
Water temperature (°C)	21	24	0	5	10	15	--	--	--	--	--
Hardness (mg/L as CaCO3)	13	14	0	10	225	1,800	--	--	--	--	--
Acid neutralizing capacity (mg/L as CaCO3)	12	13	0	82	340	1,800	--	--	--	--	--
Mesaverde Group											
Specific conductance (µS/cm)	45	133	0	225	1,390	6,000	--	--	--	--	--
pH (standard units)	42	129	0	5.3	7.4	11.5	6.5 minima (SMCL)	5	5	[1]1977-84	5.3-6.4
							8.5 maxima (SMCL)	12	17	[1]1977-83	8.6-11.5
Water temperature (°C)	45	138	0	3	10	50	--	--	--	--	--
Dissolved oxygen (mg/L)	46	57	0	0.0	1.6	10.0	--	--	--	--	--
Hardness (mg/L as CaCO3)	40	128	0	6	450	3,250	--	--	--	--	--
Acid neutralizing capacity (mg/L as CaCO3)	39	54	0	74	430	2,760	--	--	--	--	--

Table 21. Summary statistics for selected physical property water-quality data and exceedances of Colorado Department of Public Health and Environment water-quality standards for groundwater in the Upper Yampa River watershed, Colorado, 1975 through 1989 and 1998.—Continued

[No., number; Min., minimum; Max., maximum; µS/cm, microsiemens per centimeter at 25 degrees Celsius; --, no water-quality standard; SMCL, secondary maximum contaminant level; C, degrees Celsius; mg/L, milligrams per liter; CaCO₃, calcium carbonate; ne, no exceedance. Number of significant figures for individual constituents may vary among the geologic units because data are from different sources and different analytical periods. Water-quality standards are from Colorado Department of Public Health and Environment (2009b)]

Physical property (reporting units)	Total no. of wells	Total no. of samples	Total no. of censored values	Min. value	Median value	Max. value	Water-quality standard	Exceedances of water-quality standard			
								No. of wells	No. of samples	Year(s)	Range of values not meeting standard
Precambrian erathem											
Specific conductance (µS/cm)	2	2	0	103	163	223	--	--	--	--	--
pH (standard units)	2	2	0	7.1	7.6	8.0	6.5 minima (SMCL)	0	0	ne	ne
							8.5 maxima (SMCL)	0	0	ne	ne
Water temperature (°C)	2	2	0	5	6	7	--	--	--	--	--
Hardness (mg/L as CaCO₃)	1	1	0	40	40	40	--	--	--	--	--
Acid neutralizing capacity (mg/L as CaCO₃)	1	1	0	49	49	49	--	--	--	--	--
Upper Cretaceous series											
Specific conductance (µS/cm)	3	3	0	880	890	6,000	--	--	--	--	--
pH (standard units)	3	3	0	7.3	7.4	8.1	6.5 minima (SMCL)	0	0	ne	ne
							8.5 maxima (SMCL)	0	0	ne	ne
Water temperature (°C)	3	3	0	9	12	13.5	--	--	--	--	--
Hardness (mg/L as CaCO₃)	2	2	0	25	188	350	--	--	--	--	--
Acid neutralizing capacity (mg/L as CaCO₃)	1	1	0	378	378	378	--	--	--	--	--
Valley-fill deposits											
Specific conductance (µS/cm)	7	8	0	135	380	1,450	--	--	--	--	--
pH (standard units)	7	8	0	6.1	7.1	7.5	6.5 minima (SMCL)	1	1	1975	6.1
							8.5 maxima (SMCL)	0	0	ne	ne
Water temperature (°C)	7	10	0	9	13	20	--	--	--	--	--
Hardness (mg/L as CaCO₃)	7	7	0	58	170	820	--	--	--	--	--
Acid neutralizing capacity (mg/L as CaCO₃)	7	7	0	61	122	475	--	--	--	--	--
Unknown geologic units											
Specific conductance (µS/cm)	78	336	0	122	1,325	15,900	--	--	--	--	--
pH (standard units)	78	332	0	6.6	7.8	12.3	6.5 minima (SMCL)	0	0	ne	ne
							8.5 maxima (SMCL)	17	54	1975–89[1]	8.6–12.3
Water temperature (°C)	103	339	0	2	10	37	--	--	--	--	--
Dissolved oxygen (mg/L)	16	58	0	0.3	1.2	7.4	--	--	--	--	--
Hardness (mg/L as CaCO₃)	78	339	0	4.87	300	7,000	--	--	--	--	--
Acid neutralizing capacity (mg/L as CaCO₃)	53	68	0	33	372	700	--	--	--	--	--

[1]Exceedances did not occur in every year of the period of record.

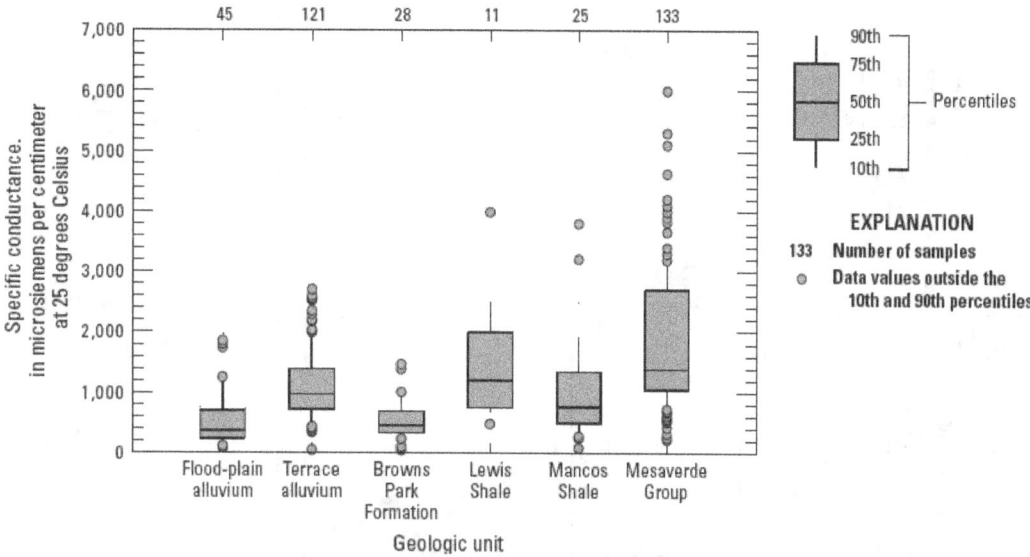

Figure 16.   Specific conductance in groundwater samples from selected geologic units in the Upper Yampa River watershed, Colorado, 1975 through 1986.

less than 0.5 mg/L (anoxic, reducing conditions) can be more susceptible to contamination from sulfate, iron, manganese, and other water-quality constituents (Jurgens and others, 2009). For the UYRW, there were insufficient data to relate concentrations of nitrate, arsenic, and other chemical constituents in groundwater to redox conditions. Less than 3 percent (23 of 816) of groundwater samples could be classified as oxic or anoxic using the redox framework devised by Jurgens and others (2009) to classify and assess redox conditions in groundwater.

Dissolved Solids and Majors Ions

Dissolved solids data were available for 564 groundwater samples collected from 191 wells (table 22). Concentrations ranged from 46 to 8,490 mg/L with a median of 812 mg/L. Among the geologic units with more than one sample, the median concentration was lowest (255 mg/L) for samples from the valley-fill deposits and highest (894 mg/L) for samples from the Mesaverde Group.

Major ion data were available for 559 groundwater samples collected from as many as 186 wells. A total of 184 samples with data for the cations calcium, magnesium, sodium, and potassium and the anions bicarbonate, chloride, fluoride, and sulfate were used to calculate water type for geologic units with multiple samples. Calcium and (or) calcium plus magnesium or a mixture of these with sodium were the dominant cations (fig. 17). Sodium was more common in groundwater from the Mesaverde Group than the other major geologic units. Bicarbonate was the dominant anion, but many samples also had sulfate, particularly those from the terrace alluvium and the Mesaverde Group (fig. 17). Calcium and magnesium can come from the dissolution of calcite

and dolomite from limey shales, limestones, and dolomitic limestones, and sodium can come from cation exchange reactions with calcium and magnesium ions (Robson and Stewart, 1990). Dissolution of carbonate minerals is likely the source of bicarbonate; sulfate may occur because of the dissolution of gypsum and oxidation of the minerals pyrite and marcasite (Robson and Stewart, 1990). Insufficient leaching of materials in soil from the lack of precipitation in a semiarid climate may cause sulfate to remain near the land surface (Hem, 1992).

Dissolved sulfate concentrations ranged from less than 1 to 4,000 mg/L with a median of 220 mg/L (table 22). For the geologic units with multiple samples, the median dissolved sulfate concentration was lowest (5.9 mg/L) for samples from the Browns Park Formation and highest (315 mg/L) for samples from the terrace alluvium. Concentrations in almost one-half (250 of 554) of the samples analyzed for dissolved sulfate were greater than the CDPHE SMCL of 250 mg/L for groundwater (table 22) (Colorado Department of Public Health and Environment, 2009b). Between about 47 and 57 percent of samples collected from the terrace alluvium, Mesaverde Group, and unknown geologic units had dissolved sulfate concentrations that did not meet the standard. Wells with concentrations exceeding the standards were commonly completed in unknown geologic units (mostly in the Yampa coal field) and terrace alluvium, about 54 and 64 percent of wells, respectively. All terrace alluvium wells were in one small area (Area A on fig. 15) in the UYRW downstream from Sage Creek.

The range in dissolved chloride concentrations of 0.5 to 5,000 mg/L was similar to that for dissolved sulfate, but the median chloride concentration of 10 mg/L was much lower than that for sulfate (table 22). For the geologic units with more than one sample, median concentrations were 6.3 mg/L or less from samples from the flood-plain alluvium, Browns Park Formation, and valley-fill deposits. Median

Table 22. Summary statistics for dissolved solids and selected major ion water-quality data and exceedances of Colorado Department of Public Health and Environment water-quality standards for groundwater in the Upper Yampa River watershed, Colorado, 1975 through 1989 and 1998

[No., number; Min., minimum; Max., maximum; --, no water-quality standard; <, less than; SMCL, secondary maximum contaminant level; MCL, maximum contaminant level; ne, no exceedance. All constituents are reported in milligrams per liter. Number of significant figures for individual constituents may vary among the geologic units because data are from different sources and different analytical periods. Water-quality standards are from Colorado Department of Public Health and Environment (2009b)]

Constituent	Total no. of wells	Total no. of samples	Total no. of censored values[1]	Min. value	Median value	Max. value	Water-quality standard	Exceedances of water-quality standard			
								No. of wells	No. of samples	Year(s)	Range of values not meeting standard
Upper Yampa River watershed											
Dissolved solids	191	564	0	46	812	8,490	--	--	--	--	--
Calcium, dissolved	186	559	0	1.3	73	2,800	--	--	--	--	--
Magnesium, dissolved	186	559	2	<0.1	33	470	--	--	--	--	--
Sodium, dissolved	185	558	0	2.2	150	1,300	--	--	--	--	--
Potassium, dissolved	186	559	0	0.3	3.8	1,500	--	--	--	--	--
Bicarbonate, unfiltered	158	208	0	0	372	3,360	--	--	--	--	--
Sulfate, dissolved	185	554	1	<1	220	4,000	250 (SMCL)	71	250	[2]1975–89	260–4,000
Chloride, dissolved	185	554	0	0.5	10	5,000	250 (SMCL)	8	15	1975–82	280–5,000
Fluoride, dissolved	184	557	0	0.1	0.4	15	4 (MCL)	4	5	[2]1975–89	4.1–15
Silica, dissolved	184	557	1	[3]0.3	11	150	--	--	--	--	--
Alluvium, flood plain											
Dissolved solids	14	15	0	46	289	1,140	--	--	--	--	--
Calcium, dissolved	14	15	0	5.6	47	80	--	--	--	--	--
Magnesium, dissolved	14	15	0	1.1	15	90	--	--	--	--	--
Sodium, dissolved	14	15	0	2.2	17	240	--	--	--	--	--
Potassium, dissolved	14	15	0	0.5	2.0	3.2	--	--	--	--	--
Bicarbonate, unfiltered	14	15	0	23	233	637	--	--	--	--	--
Sulfate, dissolved	14	15	0	3.8	49	370	250 (SMCL)	1	1	1975	370
Chloride, dissolved	14	15	0	1.8	6.3	61	250 (SMCL)	0	0	ne	ne
Fluoride, dissolved	14	15	0	0.1	0.3	2.3	4 (MCL)	0	0	ne	ne
Silica, dissolved	14	15	0	7.0	13	20	--	--	--	--	--
Alluvium, terrace											
Dissolved solids	22	44	0	205	733	2,110	--	--	--	--	--
Calcium, dissolved	22	44	0	31	96	270	--	--	--	--	--
Magnesium, dissolved	22	44	0	14	46	120	--	--	--	--	--
Sodium, dissolved	22	44	0	22	73	230	--	--	--	--	--
Potassium, dissolved	22	44	0	2.5	4.7	10	--	--	--	--	--
Bicarbonate, unfiltered	22	39	0	140	210	490	--	--	--	--	--
Sulfate, dissolved	22	44	0	42	315	1,200	250 (SMCL)	14	25	1978–79	280–1,200
Chloride, dissolved	22	44	0	8.7	42	200	250 (SMCL)	0	0	ne	ne
Fluoride, dissolved	22	44	0	0.2	0.7	1.2	4 (MCL)	0	0	ne	ne
Silica, dissolved	22	44	0	8.2	14	23	--	--	--	--	--

Table 22. Summary statistics for dissolved solids and selected major ion water-quality data and exceedances of Colorado Department of Public Health and Environment water-quality standards for groundwater in the Upper Yampa River watershed, Colorado, 1975 through 1989 and 1998.—Continued

[No, number; Min., minimum; Max., maximum; --, less than; SMCL, secondary maximum contaminant level; MCL, maximum contaminant level; ne, no exceedance. All constituents are reported in milligrams per liter. Number of significant figures for individual constituents may vary among the geologic units because data are from different sources and different analytical periods. Water-quality standards are from Colorado Department of Public Health and Environment (2009b)]

Constituent	Total no. of wells	Total no. of samples	Total no. of censored values[1]	Min. value	Median value	Max. value	Water-quality standard	Exceedances of water-quality standard — No. of wells	No. of samples	Year(s)	Range of values not meeting standard
Browns Park Formation											
Dissolved solids	7	7	0	203	329	813	--	--	--	--	--
Calcium, dissolved	7	7	0	3.7	59	87	--	--	--	--	--
Magnesium, dissolved	7	7	0	0.2	6.7	15	--	--	--	--	--
Sodium, dissolved	7	7	0	6.3	13	340	--	--	--	--	--
Potassium, dissolved	7	7	0	0.7	2.0	9.3	--	--	--	--	--
Bicarbonate, unfiltered	7	7	0	180	250	720	--	--	--	--	--
Sulfate, dissolved	7	7	0	4.6	5.9	110	250 (SMCL)	0	0	ne	ne
Chloride, dissolved	7	7	0	1.9	4.1	91	250 (SMCL)	0	0	ne	ne
Fluoride, dissolved	7	7	0	0.1	0.2	15	4 (MCL)	1	1	1978	15
Silica, dissolved	7	7	0	10	18	60	--	--	--	--	--
Curtis Formation of San Rafael Group											
Dissolved solids	1	1	0	263	263	263	--	--	--	--	--
Calcium, dissolved	1	1	0	51	51	51	--	--	--	--	--
Magnesium, dissolved	1	1	0	13	13	13	--	--	--	--	--
Sodium, dissolved	1	1	0	27	27	27	--	--	--	--	--
Potassium, dissolved	1	1	0	1.3	1.3	1.3	--	--	--	--	--
Bicarbonate, unfiltered	1	1	0	260	260	260	--	--	--	--	--
Sulfate, dissolved	1	1	0	23	23	23	250 (SMCL)	0	0	ne	ne
Chloride, dissolved	1	1	0	1.5	1.5	1.5	250 (SMCL)	0	0	ne	ne
Fluoride, dissolved	1	1	0	0.1	0.1	0.1	4 (MCL)	0	0	ne	ne
Silica, dissolved	1	1	0	17	17	17	--	--	--	--	--
Eocene series											
Dissolved solids	1	1	0	476	476	476	--	--	--	--	--
Calcium, dissolved	1	1	0	3.8	3.8	3.8	--	--	--	--	--
Magnesium, dissolved	1	1	0	0.6	0.6	0.6	--	--	--	--	--
Sodium, dissolved	1	1	0	190	190	190	--	--	--	--	--
Potassium, dissolved	1	1	0	1.3	1.3	1.3	--	--	--	--	--
Bicarbonate, unfiltered	1	1	0	446	446	446	--	--	--	--	--
Sulfate, dissolved	1	1	0	42	42	42	250 (SMCL)	0	0	ne	ne
Chloride, dissolved	1	1	0	7.3	7.3	7.3	250 (SMCL)	0	0	ne	ne
Fluoride, dissolved	1	1	0	2.7	2.7	2.7	4 (MCL)	0	0	ne	ne
Silica, dissolved	1	1	0	8.4	8.4	8.4	--	--	--	--	--

Table 22. Summary statistics for dissolved solids and selected major ion water-quality data and exceedances of Colorado Department of Public Health and Environment water-quality standards for groundwater in the Upper Yampa River watershed, Colorado, 1975 through 1989 and 1998.—Continued

[No., number; Min., minimum; Max., maximum; --, no value; <, less than; SMCL, secondary maximum contaminant level; MCL, maximum contaminant level; ne, no exceedance. All constituents are reported in milligrams per liter. Number of significant figures for individual constituents may vary among the geologic units because data are from different sources and different analytical periods. Water-quality standards are from Colorado Department of Public Health and Environment (2009b)]

Constituent	Total no. of wells	Total no. of samples	Total no. of censored values[1]	Min. value	Median value	Max. value	Water-quality standard	Exceedances of water-quality standard			
								No. of wells	No. of samples	Year(s)	Range of values not meeting standard
Fort Union Formation											
Dissolved solids	1	1	0	717	717	717	--	--	--	--	--
Calcium, dissolved	1	1	0	100	100	100	--	--	--	--	--
Magnesium, dissolved	1	1	0	34	34	34	--	--	--	--	--
Sodium, dissolved	1	1	0	61	61	61	--	--	--	--	--
Potassium, dissolved	1	1	0	4.3	4.3	4.3	--	--	--	--	--
Bicarbonate, unfiltered	1	1	0	91	91	91	--	--	--	--	--
Sulfate, dissolved	1	1	0	180	180	180	250 (SMCL)	0	0	ne	ne
Chloride, dissolved	1	1	0	95	95	95	250 (SMCL)	0	0	ne	ne
Fluoride, dissolved	1	1	0	0.3	0.3	0.3	4 (MCL)	0	0	ne	ne
Silica, dissolved	1	1	0	32	32	32	--	--	--	--	--
Lewis Shale											
Dissolved solids	6	6	0	252	743	1,360	--	--	--	--	--
Calcium, dissolved	6	6	0	4.3	90	160	--	--	--	--	--
Magnesium, dissolved	6	6	0	3.1	35	67	--	--	--	--	--
Sodium, dissolved	5	5	0	13	63	570	--	--	--	--	--
Potassium, dissolved	6	6	0	0.8	2.0	4	--	--	--	--	--
Bicarbonate, unfiltered	5	5	0	230	311	1,450	--	--	--	--	--
Sulfate, dissolved	6	6	0	30	118	290	250 (SMCL)	1	1	1975	290
Chloride, dissolved	6	6	0	3.4	34	83	250 (SMCL)	0	0	ne	ne
Fluoride, dissolved	6	6	0	0.2	0.4	3	4 (MCL)	0	0	ne	ne
Silica, dissolved	6	6	0	8.8	12	22	--	--	--	--	--
Mancos Shale											
Dissolved solids	12	13	0	121	689	3,290	--	--	--	--	--
Calcium, dissolved	12	13	0	2.2	65	360	--	--	--	--	--
Magnesium, dissolved	12	13	0	1.2	24	210	--	--	--	--	--
Sodium, dissolved	12	13	0	5.6	94	1,000	--	--	--	--	--
Potassium, dissolved	12	13	0	1.1	2.6	5.6	--	--	--	--	--
Bicarbonate, unfiltered	12	13	0	100	415	2,200	--	--	--	--	--
Sulfate, dissolved	12	13	0	7.7	100	1,700	250 (SMCL)	2	2	1975, 1978	420, 1,700
Chloride, dissolved	12	13	0	1.8	12	350	250 (SMCL)	3	3	1975, 1978	300–350
Fluoride, dissolved	12	13	0	0.1	0.3	5.1	4 (MCL)	2	2	1975	4.8, 5.1
Silica, dissolved	12	13	0	6.3	13	150	--	--	--	--	--

Table 22. Summary statistics for dissolved solids and selected major ion water-quality data and exceedances of Colorado Department of Public Health and Environment water-quality standards for groundwater in the Upper Yampa River watershed, Colorado, 1975 through 1989 and 1998.—Continued

[No., number; Min., minimum; Max., maximum; --, no water-quality standard; <, less than; SMCL, secondary maximum contaminant level; MCL, maximum contaminant level; ne, no exceedance. All constituents are reported in milligrams per liter. Number of significant figures for individual constituents may vary among the geologic units because data are from different sources and different analytical periods. Water-quality standards are from Colorado Department of Public Health and Environment (2009b)]

Constituent	Total no. of wells	Total no. of samples	Total no. of censored values[1]	Min. value	Median value	Max. value	Water-quality standard	Exceedances of water-quality standard			
								No. of wells	No. of samples	Year(s)	Range of values not meeting standard
Mesaverde Group											
Dissolved solids	40	128	0	133	894	6,560	--	--	--	--	--
Calcium, dissolved	41	129	0	1.8	89	960	--	--	--	--	--
Magnesium, dissolved	41	129	2	<0.1	53	420	--	--	--	--	--
Sodium, dissolved	41	129	0	3.7	190	930	--	--	--	--	--
Potassium, dissolved	41	129	0	48	6.3	1,500	--	--	--	--	--
Bicarbonate, unfiltered	39	55	0	90	518	3,360	--	--	--	--	--
Sulfate, dissolved	41	126	0	1.2	260	2,900	250 (SMCL)	15	64	[2]1975–86	260–2,900
Chloride, dissolved	41	126	0	1.2	9.5	1,900	250 (SMCL)	2	3	[2]1976–80	280–1,900
Fluoride, dissolved	40	128	0	0.1	0.4	3.1	4 (MCL)	0	0	ne	ne
Silica, dissolved	41	129	0	0.3	10	35	--	--	--	--	--
Precambrian erathem											
Dissolved solids	1	1	0	64	64	64	--	--	--	--	--
Calcium, dissolved	1	1	0	9.2	9.2	9.2	--	--	--	--	--
Magnesium, dissolved	1	1	0	4.2	4.2	4.2	--	--	--	--	--
Sodium, dissolved	1	1	0	3.2	3.2	3.2	--	--	--	--	--
Potassium, dissolved	1	1	0	1.4	1.4	1.4	--	--	--	--	--
Bicarbonate, unfiltered	1	1	0	60	60	60	--	--	--	--	--
Sulfate, dissolved	1	1	0	2.7	2.7	2.7	250 (SMCL)	0	0	ne	ne
Chloride, dissolved	1	1	0	4.2	4.2	4.2	250 (SMCL)	0	0	ne	ne
Fluoride, dissolved	1	1	0	0.2	0.2	0.2	4 (MCL)	0	0	ne	ne
Silica, dissolved	1	1	0	8.6	8.6	8.6	--	--	--	--	--
Upper Cretaceous series											
Dissolved solids	1	1	0	569	569	569	--	--	--	--	--
Calcium, dissolved	1	1	0	87	87	87	--	--	--	--	--
Magnesium, dissolved	1	1	0	33	33	33	--	--	--	--	--
Sodium, dissolved	1	1	0	70	70	70	--	--	--	--	--
Potassium, dissolved	1	1	0	2.6	2.6	2.6	--	--	--	--	--
Bicarbonate, unfiltered	1	1	0	461	461	461	--	--	--	--	--
Sulfate, dissolved	1	1	0	110	110	110	250 (SMCL)	0	0	ne	ne
Chloride, dissolved	1	1	0	19	19	19	250 (SMCL)	0	0	ne	ne
Fluoride, dissolved	1	1	0	0.3	0.3	0.3	4 (MCL)	0	0	ne	ne
Silica, dissolved	1	1	0	11	11	11	--	--	--	--	--

Table 22. Summary statistics for dissolved solids and selected major ion water-quality data and exceedances of Colorado Department of Public Health and Environment water-quality standards for groundwater in the Upper Yampa River watershed, Colorado, 1975 through 1989 and 1998.—Continued

[No., number; Min., minimum; Max., maximum; --, no exceedance; <, less than; SMCL, secondary maximum contaminant level; MCL, maximum contaminant level; ne, no exceedance. All constituents are reported in milligrams per liter. Number of significant figures for individual constituents may vary among the geologic units because data are from different sources and different analytical periods. Water-quality standards are from Colorado Department of Public Health and Environment (2009b)]

Constituent	Total no. of wells	Total no. of samples	Total no. of censored values[1]	Min. value	Median value	Max. value	Water-quality standard	Exceedances of water-quality standard			
								No. of wells	No. of samples	Year(s)	Range of values not meeting standard
Valley-fill deposits											
Dissolved solids	7	7	0	82	255	1,050	--	--	--	--	--
Calcium, dissolved	7	7	0	16	49	200	--	--	--	--	--
Magnesium, dissolved	7	7	0	4.5	11	77	--	--	--	--	--
Sodium, dissolved	7	7	0	3.6	23	34	--	--	--	--	--
Potassium, dissolved	7	7	0	0.3	1.7	5	--	--	--	--	--
Bicarbonate, unfiltered	7	7	0	74	149	557	--	--	--	--	--
Sulfate, dissolved	7	7	0	1.7	34	420	250 (SMCL)	1	1	1975	420
Chloride, dissolved	7	7	0	0.5	4	21	250 (SMCL)	0	0	ne	ne
Fluoride, dissolved	7	7	0	0.1	0.2	0.7	4 (MCL)	0	0	ne	ne
Silica, dissolved	7	7	0	11	14	26	--	--	--	--	--
Unknown geologic units											
Dissolved solids	78	339	0	89	845	8,490	--	--	--	--	--
Calcium, dissolved	72	333	0	1.3	66	2,800	--	--	--	--	--
Magnesium, dissolved	72	333	0	0.3	26	470	--	--	--	--	--
Sodium, dissolved	72	333	0	9.9	160	1,300	--	--	--	--	--
Potassium, dissolved	72	333	0	0.8	3.4	170	--	--	--	--	--
Bicarbonate, unfiltered	47	62	0	0	470	732	--	--	--	--	--
Sulfate, dissolved	71	331	1	<1	240	4,000	250 (SMCL)	38	156	[2]1975–89	260–4,000
Chloride, dissolved	72	331	0	1.4	9.1	5,000	250 (SMCL)	3	9	[2]1975–82	350–5,000
Fluoride, dissolved	72	332	0	0.1	0.4	4.3	4 (MCL)	1	2	1988, 1989	4.1, 4.3
Silica, dissolved	70	331	1	[3]1	10	38	--	--	--	--	--

[1] Censored values can be expressed as values less than the laboratory reporting level.

[2] Exceedances did not occur in every year of the period of record.

[3] The minimum censored value is greater than the minimum detected value that is shown.

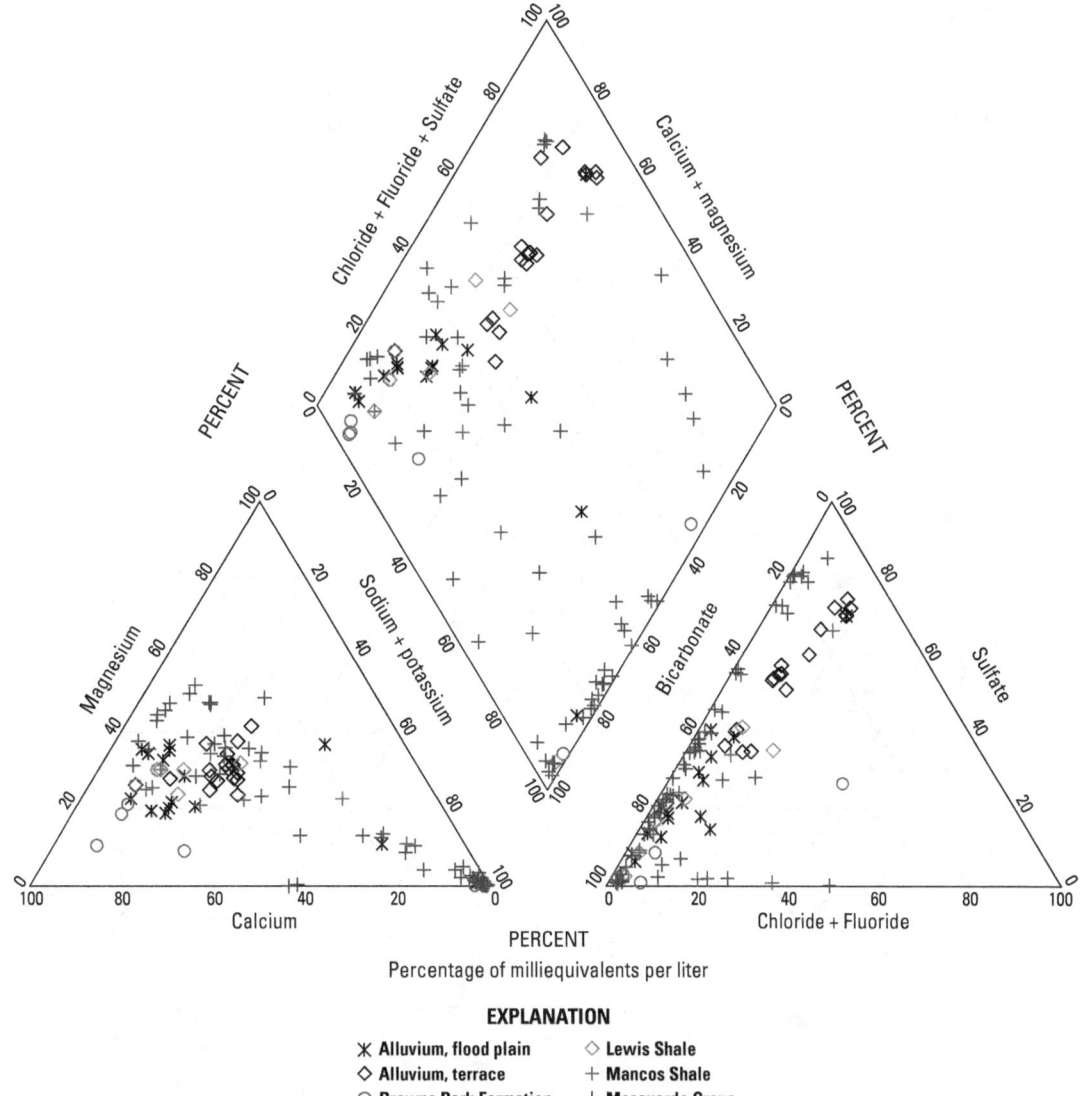

Figure 17. Major cation and anion percentages and water type for groundwater samples from selected geologic units in the Upper Yampa River watershed, Colorado, 1975 through 1984.

concentrations of 12 and 34 mg/L in the Mancos and Lewis Shales, respectively, probably reflect the marine origin of the formations. Concentrations of dissolved chloride in about 2.7 percent (15 of 554) of samples did not meet the CDPHE SMCL of 250 mg/L for groundwater (table 22) (Colorado Department of Public Health and Environment, 2009b). Most (9 of 15) exceedances were in samples from wells with no geologic unit description; three exceedances each were in samples from the Mancos Shale and Mesaverde Group.

Dissolved fluoride concentrations generally were much lower than concentrations for the other major ions studied. Concentrations of dissolved fluoride ranged from 0.1 to 15 mg/L, and the median was 0.4 mg/L (table 22). Five of 557 dissolved fluoride concentrations did not meet the CDPHE MCL of 4 mg/L for groundwater (table 22) (Colorado Department of Public Health and Environment, 2009b).

Nutrients

Nutrient data were available for 494 groundwater samples (table 19) collected from 189 wells. Dissolved nitrate plus nitrite data were the most common (473 samples), and total unfiltered phosphorus data were the least common (42 samples) (table 23). Data for dissolved nitrite only were available for wells in the Mesaverde Group and unknown geologic units. Most (112 of 142) dissolved nitrite concentrations were less than laboratory detection levels (table 23). All concentrations were 0.26 mg/L or less, well below the CDPHE MCL of 1 mg/L of nitrite in groundwater and the agricultural-use standard for livestock watering of 10 mg/L (table 23) (Colorado Department of Public Health and Environment, 2009b). Data for dissolved nitrate plus nitrite were most common for samples from wells with no geologic unit description

Table 23.   Summary statistics for selected nutrient water-quality data and exceedances of Colorado Department of Public Health and Environment water-quality standards for groundwater in the Upper Yampa River watershed, Colorado, 1975 through 1989 and 1998.

[No., number; Min., minimum; Max., maximum; N, nitrogen; <, less than; MCL, maximum contaminant level; Ag, Agriculture; ne, no exceedance; E, estimated; --, no water-quality standard; P, phosphorus. All constituents are reported in milligrams per liter. Number of significant figures for individual constituents may vary among the geologic units because data are from different sources and different analytical periods. Water-quality standards are from Colorado Department of Public Health and Environment (2009b)]

Constituent	Total no. of wells	Total no. of samples	Total no. of censored values[1]	Min. value	Median value	Max. value	Water-quality standard	Exceedances of water-quality standard				
								Standard	No. of wells	No. of samples	Year(s)	Range of values not meeting standard
Upper Yampa River watershed												
Nitrite, dissolved, as N	27	142	112	<0.01	<0.01	0.26	1 (MCL), 10 (Ag)	ne	0	0	ne	ne
Nitrate plus nitrite, dissolved, as N	182	467	128	<0.01	E 0.11	37	10 (MCL), 100 (Ag)	MCL	13	21	[3]1975–89	12–37
Total ammonia, dissolved, as N	29	144	0	0.03	0.45	5.6	--	--	--	--	--	--
Total phosphorus, dissolved, as P	68	278	54	<0.01	0.02	0.76	--	--	--	--	--	--
Total phosphorus, unfiltered, as P	42	42	10	<0.01	0.02	0.33	--	--	--	--	--	--
Orthophosphate, dissolved, as P	180	375	134	<0.01	E 0.03	2.4	--	--	--	--	--	--
Alluvium, flood plain												
Nitrate plus nitrite, dissolved, as N	14	15	0	0.01	0.81	18	10 (MCL)[2], 100 (Ag)	MCL	1	1	1976	18
Total phosphorus, unfiltered, as P	8	8	3	<0.01	0.02	0.08	--	--	--	--	--	--
Orthophosphate, dissolved, as P	14	15	5	<0.01	0.03	0.09	--	--	--	--	--	--
Alluvium, terrace												
Nitrate plus nitrite, dissolved, as N	22	45	0	0.11	1.5	20	10 (MCL)[2], 100 (Ag)	MCL	1	3	1979	18, 20
Total ammonia, dissolved, as N	1	1	0	1.3	1.3	1.3	--	--	--	--	--	--
Total phosphorus, dissolved, as P	22	44	0	0.01	0.05	0.23	--	--	--	--	--	--
Orthophosphate, dissolved, as P	22	26	7	<0.01	0.04	0.43	--	--	--	--	--	--
Browns Park Formation												
Nitrate plus nitrite, dissolved, as N	7	7	0	0.04	0.34	13	10 (MCL)[2], 100 (Ag)	MCL	1	1	1978	13
Orthophosphate, dissolved, as P	7	7	1	<0.01	0.03	0.31	--	--	--	--	--	--
Curtis Formation of San Rafael Group												
Nitrate plus nitrite, dissolved, as N	1	1	0	0.13	0.13	0.13	10 (MCL)[2], 100 (Ag)	ne	0	0	ne	ne
Orthophosphate, dissolved, as P	1	1	1	<0.01	<0.01	<0.01	--	--	--	--	--	--
Eocene series												
Nitrate plus nitrite, dissolved, as N	1	1	0	0.02	0.02	0.02	10 (MCL)[2], 100 (Ag)	ne	0	0	ne	ne
Total phosphorus, unfiltered, as P	1	1	0	0.03	0.03	0.03	--	--	--	--	--	--
Orthophosphate, dissolved, as P	1	1	0	0.09	0.09	0.09	--	--	--	--	--	--
Fort Union Formation												
Nitrate plus nitrite, dissolved, as N	1	1	0	37	37	37	10 (MCL)[2], 100 (Ag)	MCL	1	1	1975	37
Orthophosphate, dissolved, as P	1	1	0	0.49	0.49	0.49	--	--	--	--	--	--
Lewis Shale												
Nitrate plus nitrite, dissolved, as N	5	5	0	0.02	1.1	15	10 (MCL)[2], 100 (Ag)	MCL	2	2	1975	12, 15
Total phosphorus, dissolved, as P	1	1	1	<0.03	<0.03	<0.03	--	--	--	--	--	--
Total phosphorus, unfiltered, as P	5	5	1	<0.01	0.02	0.22	--	--	--	--	--	--
Orthophosphate, dissolved, as P	5	5	3	<0.01	<0.01	0.43	--	--	--	--	--	--
Mancos Shale												
Nitrate plus nitrite, dissolved, as N	12	13	0	0.01	0.34	17	10 (MCL)[2], 100 (Ag)	MCL	1	1	1978	17
Total phosphorus, unfiltered, as P	2	2	0	0.01	0.06	0.12	--	--	--	--	--	--
Orthophosphate, dissolved, as P	12	13	5	<0.01	0.03	0.12	--	--	--	--	--	--

Table 23. Summary statistics for selected nutrient water-quality data and exceedances of Colorado Department of Public Health and Environment water-quality standards for groundwater in the Upper Yampa River watershed, Colorado, 1975 through 1989 and 1998.—Continued

[No., number; Min., minimum; Max., maximum; N, nitrogen; <, less than; MCL, maximum contaminant level; Ag, Agriculture; ne, no exceedance; E, estimated; --, no water-quality standard. All constituents are reported in milligrams per liter. Number of significant figures for individual constituents may vary among the geologic units because data are from different sources and different analytical periods. Water-quality standards are from Colorado Department of Public Health and Environment (2009b)]

Constituent	Total no. of wells	Total no. of samples	Total no. of censored values[1]	Min. value	Median value	Max. value	Water-quality standard	Exceedances of water-quality standard				
								Standard	No. of wells	No. of samples	Year(s)	Range of values not meeting standard
Mesaverde Group												
Nitrite, dissolved, as N	3	3	2	<0.01	<0.01	0.03	1 (MCL), 10 (Ag)	ne	0	0	ne	ne
Nitrate plus nitrite, dissolved, as N	40	120	13	<0.1	0.12	14	10 (MCL)[2], 100 (Ag)	MCL	1	1	1978	14
Total ammonia, dissolved, as N	4	4	0	0.11	0.8	1.8	--	--	--	--	--	--
Total phosphorus, dissolved, as P	10	44	9	<0.1	0.01	0.07	--	--	--	--	--	--
Total phosphorus, unfiltered, as P	9	9	2	<0.01	0.03	0.33	--	--	--	--	--	--
Orthophosphate, dissolved, as P	40	82	18	<0.01	0.06	0.98	--	--	--	--	--	--
Precambrian erathem												
Nitrate plus nitrite, dissolved, as N	1	1	1	<0.1	<0.1	<0.1	10 (MCL)[2], 100 (Ag)	ne	0	0	ne	ne
Orthophosphate, dissolved, as P	1	1	0	0.06	0.06	0.06	--	--	--	--	--	--
Upper Cretaceous series												
Nitrate plus nitrite, dissolved, as N	1	1	0	1.9	1.9	1.9	10 (MCL)[2], 100 (Ag)	ne	0	0	ne	ne
Orthophosphate, dissolved, as P	1	1	0	0.03	0.03	0.03	--	--	--	--	--	--
Valley-fill deposits												
Nitrate plus nitrite, dissolved, as N	7	7	0	0.01	0.32	8.8	10 (MCL)[2], 100 (Ag)	ne	0	0	ne	ne
Total phosphorus, unfiltered, as P	4	4	0	0.03	0.02	0.09	--	--	--	--	--	--
Orthophosphate, dissolved, as P	7	7	3	<0.01	0.03	0.12	--	--	--	--	--	--
Unknown geologic units												
Nitrite, dissolved, as N	103	139	110	<0.01	<0.01	0.26	1 (MCL), 10 (Ag)	ne	0	0	ne	ne
Nitrate plus nitrite, dissolved, as N	70	250	114	<0.01	0.06	27	10 (MCL)[2], 100 (Ag)	MCL	5	11	1988–89	13–27
Total ammonia, dissolved, as N	103	139	0	0.03	0.45	5.6	--	--	--	--	--	--
Total phosphorus, dissolved, as P	103	189	45	<0.01	0.02	0.76	--	--	--	--	--	--
Total phosphorus, unfiltered, as P	13	13	4	<0.01	0.02	0.12	--	--	--	--	--	--
Orthophosphate, dissolved, as P	68	215	91	<0.01	0.03	2.4	--	--	--	--	--	--

[1]Censored values can be expressed as values less than the laboratory reporting level.

[2]The MCL of 10 mg/L is for nitrate, as N.

[3]Exceedances did not occur in every year of the period of record.

(250 samples) and the Mesaverde Group (120 samples) (table 23). Median concentrations were 0.81 mg/L or less for all geologic units with two of more samples except for the terrace alluvium (1.5 mg/L) and Lewis Shale (1.1 mg/L). About 4.5 percent (21 of 467) of samples with nitrate plus nitrite data had concentrations that exceeded the CDPHE MCL standard of 10 mg/L for nitrate (table 23) (Colorado Department of Public Health and Environment, 2009b). About one-half of the concentrations greater than 10 mg/L were detected in samples collected before 1980, and others were collected during 1988 and 1989 (table 23). About 24 percent of the wells with sample concentrations exceeding the standard were in the Yampa coal field. No nitrate plus nitrite concentrations exceeded the agricultural-use standard for livestock watering of 100 mg/L.

Data for dissolved total phosphorus were available for well samples with no aquifer or geologic unit description (189 samples) and those from the terrace alluvium, Mesaverde Group (44 samples each) and Lewis Shale (1 sample) (table 23). Unfiltered total phosphorus data were available for only 42 samples from the flood-plain alluvium, Eocene Series, Lewis and Mancos Shales, Mesaverde Group, valley-fill deposits, and unknown geologic units (table 23). Concentrations of dissolved and unfiltered total phosphorus were 0.33 mg/L or less, except for a concentration of 0.76 mg/L in a sample from a well (site 355) in an unknown geologic unit. About 95 percent of the concentrations were less than 0.1 mg/L. The CDPHE has not established water-quality standards for dissolved and unfiltered total phosphorus in groundwater.

Trace Elements

Trace-element data were available for 609 samples from 190 wells (table 19). Samples had concentration data for the dissolved constituents and total recoverable iron, listed in table 24 and discussed in this report, and concentration data for total recoverable copper, dissolved lithium, unfiltered and total recoverable manganese, and dissolved vanadium, which are not listed in table 24 and are not discussed in this report. More than 500 samples had concentration data for dissolved boron, iron, manganese, selenium, and zinc; the fewest samples (22) were available for dissolved antimony (table 24).

More than 80 percent of the data for dissolved antimony, arsenic, beryllium, cadmium, chromium, cobalt, copper, lead, mercury, molybdenum, nickel, selenium, and silver were less than or equal to laboratory detection levels (table 24). Low concentrations are generally due to small occurrences in soils and rocks and (or) low solubility of the elements (Driver and others, 1984). All concentrations of dissolved antimony, barium, chromium, cobalt, mercury, nickel, and silver met CDPHE HH standards, MCLs, SMCLs, and (or) agricultural-use standards for groundwater (table 25). Concentrations of dissolved arsenic, beryllium, copper, lead, molybdenum, and selenium in one or two samples for each constituent were greater than HH standards, MCLs, and (or) agricultural-use standards. Only one sample collected from the flood-plain

alluvium at site 238 had an arsenic concentration greater than the MCL for arsenic of 10 µg/L. The MCL for dissolved beryllium was exceeded in two samples from one well (site 329) in an unknown geologic unit. Only one sample collected from site 358 in the valley-fill deposit exceeded the agricultural-use standard for copper of 200 µg/L. The HH and agricultural-use standards for dissolved lead were exceeded in one sample from the Mesaverde Group at site 286 near the mouth of Middle Creek. The HH standard for dissolved molybdenum was exceeded in one sample from the Mancos Shale at site 467 just northwest of Steamboat Springs. Only one sample collected from site 378 in the terrace alluvium exceeded the agricultural-use standard for selenium of 20 µg/L. The MCL for dissolved cadmium was exceeded in less than 4 percent (13 of 329) of samples; 3 samples from the Mesaverde Group near Foidel Creek and 10 samples from unknown geologic units near Fish Creek and in the area between Grassy and Sage Creeks in the Yampa coal field. Four of the samples had concentrations that exceeded the cadmium agricultural-use standard of 10 µg/L. All dissolved cadmium samples but one were from wells in the Yampa coal field.

The highest concentrations (3,600 µg/L or more) were detected for dissolved aluminum and boron, dissolved and total recoverable iron, and dissolved manganese, strontium, and zinc (table 24). All dissolved aluminum concentrations were less than the CDPHE agricultural-use standard for livestock watering of 5,000 µg/L. The CDPHE has not established a drinking-water standard for dissolved aluminum in groundwater. Boron concentrations in about 13 percent (70 of 539) of samples exceeded the agricultural-use standard of 750 µg/L. Almost 84 percent of these samples were collected from the terrace alluvium. All zinc concentrations were less than the CDPHE SMCL of 5,000 µg/L; fewer than one percent of samples had zinc concentrations that exceeded the agricultural-use standard of 2,000 µg/L. The CDPHE has not established groundwater standards for total recoverable iron and dissolved strontium. Iron and manganese are discussed in the following paragraphs.

Dissolved iron data were available for 574 samples from every lithology listed in table 20. Most samples were collected from the unknown geologic units (329 samples), Mesaverde Group (127 samples) and terrace alluvium (65 samples) (table 24). The highest median concentration per geologic unit with more than one sample was for samples from the valley-fill deposits (140 µg/L), and the lowest (less than 10 µg/L) was for samples from the Browns Park Formation. Individual samples with concentrations greater than 2,000 µg/L were collected from unknown geologic units (5 samples) and Mesaverde Group (3 samples); one sample each was collected from the flood-plain alluvium, Mancos Shale, and valley-fill deposits. The CDPHE SMCL for dissolved iron of 300 µg/L was exceeded in about 10 percent (60 of 574) of samples collected (table 24). Samples with exceedances were most commonly collected from wells in the Mancos Shale, valley-fill deposits, and unknown geologic units, between 33 to 43 percent of wells for each grouping. Concentrations exceeding the

Table 24. Summary statistics for selected trace element water-quality data and exceedances of Colorado Department of Public Health and Environment water-quality standards for groundwater in the Upper Yampa River watershed, Colorado, 1975 through 1989 and 1998.

[No, number; Min, minimum; Max, maximum; <, less than; Ag, Agriculture; ne, no exceedance; nc, not computed; MCL, maximum contaminant level; SMCL, secondary maximum contaminant level; --, no water-quality standard; HH, human health. All constituents are reported in micrograms per liter. Number of significant figures for individual constituents may vary among the geologic units because data are from different sources and analytical periods. Water-quality standards are from Colorado Department of Public Health and Environment (2009b)]

Constituent	Total no. of wells	Total no. of samples	Total no. of censored values[1]	Min. value	Median value	Max. value	Water-quality standard	Exceedances of water-quality standard			
								No. of wells	No. of samples	Year(s)	Range of values not meeting standard
Upper Yampa River watershed											
Aluminum, dissolved	68	327	166	<5	8	4,500	5,000 (Ag)	0	0	ne	ne
Antimony, dissolved	22	22	18	<1	nc	1	6 (MCL)	0	0	ne	ne
Arsenic, dissolved	176	446	248	<1	<1	160	10 (MCL), 100 (Ag)	1	1	1975	160
Barium, dissolved	24	215	0	14	69	680	2,000 (MCL)	0	0	ne	ne
Beryllium, dissolved	24	215	207	<0.5	nc	6	4 (MCL)	1	2	1988, 1989	4.5, 6.0
							100 (Ag)	0	0	ne	ne
Boron, dissolved	128	539	4	[2]10	190	4,700	750 (Ag)	24	70	1979–81	760–4,700
Cadmium, dissolved	96	329	284	<1	nc	23	5 (MCL)	13	13	1977, 1988–89	6–23
							10 (Ag)	4	4	1977, 1989	11–23
Chromium, dissolved	48	251	234	<5	nc	<25	100 (MCL, Ag)	0	0	ne	ne
Cobalt, dissolved	33	224	209	<2	nc	23	50 (Ag)	0	0	ne	ne
Copper, dissolved	78	273	241	<2	nc	300	1,000 (SMCL)	0	0	ne	ne
							200 (Ag)	1	1	1975	300
Iron, dissolved	186	574	69	<3	40	39,000	300 (SMCL)	45	60	[3]1975–89	340–39,000
							5,000 (Ag)	2	3	1980, 1981	5,100–39,000
Iron, total recoverable	42	105	0	70	2,400	590,000		--	--	--	--
Lead, dissolved	71	313	296	<1	nc	120	50 (HH), 100 (Ag)	1	1	1977	120
Manganese, dissolved	183	548	57	<1	60	3,600	50 (SMCL)	90	284	[3]1975–89	55–3,600
							200 (Ag)	44	141	[3]1975–89	210–3,600
Mercury, dissolved	140	215	201	<0.1	nc	2	10 (Ag)	0	0	ne	ne
Molybdenum, dissolved	135	395	328	<1	nc	74	35 (HH)	1	1	1978	74
Nickel, dissolved	49	242	230	<2	nc	50	100 (HH), 200 (Ag)	0	0	ne	ne
Selenium, dissolved	184	594	439	<1	<1	27	50 (MCL)	0	0	ne	ne
							20 (Ag)	1	1	1979	27
Silver, dissolved	25	216	189	<1	nc	<30	50 (HH)	0	0	ne	ne
Strontium, dissolved	50	241	0	7	560	10,000	--	--	--	--	--
Zinc, dissolved	156	514	138	<3	20	4,500	5,000 (SMCL)	0	0	ne	ne
							2,000 (Ag)	3	4	1975, 1978	2,200–4,500
Alluvium, flood plain											
Aluminum, dissolved	1	1	0	140	140	140	5,000 (Ag)	0	0	ne	ne
Antimony, dissolved	2	2	2	<1	<1	<1	6 (MCL)	0	0	ne	ne
Arsenic, dissolved	13	14	8	<1	<1	160	10 (MCL), 100 (Ag)	1	1	1975	160
Boron, dissolved	5	5	2	<20	40	700	750 (Ag)	0	0	ne	ne
Cadmium, dissolved	6	6	6	<2	<2	<2	5 (MCL), 10 (Ag)	0	0	ne	ne
Cobalt, dissolved	2	2	2	<2	<2	<2	50 (Ag)	0	0	ne	ne
Copper, dissolved	4	4	0	12	20	40	1,000 (SMCL), 200 (Ag)	0	0	ne	ne
Iron, dissolved	14	15	2	<10	40	2,500	300 (SMCL)	3	3	[3]1975–80	340–2,500
							5,000 (Ag)	0	0	ne	ne

Table 24.   Summary statistics for selected trace element water-quality data and exceedances of Colorado Department of Public Health and Environment water-quality standards for groundwater in the Upper Yampa River watershed, Colorado, 1975 through 1989 and 1998.—Continued

[No., number; Min., minimum; Max., maximum; <, less than; Ag, Agriculture; ne, no exceedance; nc, not computed; MCL, maximum contaminant level; SMCL, secondary maximum contaminant level; --, no water-quality standard; HH, human health. All constituents are reported in micrograms per liter. Number of significant figures for individual constituents may vary among the geologic units because data are from different sources and analytical periods. Water-quality standards are from Colorado Department of Public Health and Environment (2009b)]

Constituent	Total no. of wells	Total no. of samples	Total no. of censored values[1]	Min. value	Median value	Max. value	Water-quality standard	Exceedances of water-quality standard — No. of wells	No. of samples	Year(s)	Range of values not meeting standard
Alluvium, flood plain—Continued											
Iron, total recoverable	1	1	0	12,000	12,000	12,000	--	--	--	--	--
Lead, dissolved	3	3	3	<2	<2	<2	50 (HH), 100 (Ag)	0	0	ne	ne
Manganese, dissolved	14	15	6	<10	30	620	50 (SMCL); 200 (Ag)	6; 3	7; 3	[3]1975–80; 1975, 1978	60–620; 450–620
Mercury, dissolved	13	13	13	<0.1	nc	<0.5	10 (Ag)	0	0	ne	ne
Molybdenum, dissolved	13	13	8	<1	<1	15	35 (HH)	0	0	ne	ne
Nickel, dissolved	2	2	2	<2	<2	<2	100 (HH), 200 (Ag)	0	0	ne	ne
Selenium, dissolved	13	14	10	<1	nc	2	50 (MCL), 20 (Ag)	0	0	ne	ne
Strontium, dissolved	4	4	0	110	150	410	--	--	--	--	--
Zinc, dissolved	13	13	0	20	80	1,500	5,000 (SMCL), 2,000 (Ag)	0	0	ne	ne
Alluvium, terrace											
Arsenic, dissolved	22	22	1	<1	1	3	10 (MCL), 100 (Ag)	0	0	ne	ne
Boron, dissolved	22	88	0	170	1,300	4,700	750 (Ag)	20	65	1978, 1979	800–4,700
Cadmium, dissolved	15	15	15	<2	<2	<2	5 (MCL), 10 (Ag)	0	0	ne	ne
Chromium, dissolved	22	34	20	<5	10	20	100 (MCL, Ag)	0	0	ne	ne
Copper, dissolved	1	1	1	<2	<2	<2	1,000 (SMCL), 200 (Ag)	0	0	ne	ne
Iron, dissolved	22	65	32	<10	20	1,700	300 (SMCL); 5,000 (Ag)	2; 0	2; 0	1979; ne	360, 1,700; ne
Manganese, dissolved	22	67	7	<10	70	3,600	50 (SMCL); 200 (Ag)	19; 10	38; 13	1978, 1979; 1978, 1979	60–3,600; 210–3,600
Mercury, dissolved	19	19	19	<0.1	<0.1	<0.1	10 (Ag)	0	0	ne	ne
Selenium, dissolved	22	81	5	<1	4	27	50 (MCL); 20 (Ag)	0; 1	0; 1	ne; 1979	ne; 27
Zinc, dissolved	22	66	20	<3	15	40	5,000 (SMCL), 2,000 (Ag)	0	0	ne	ne
Browns Park Formation											
Arsenic, dissolved	7	7	2	<1	1	3	10 (MCL), 100 (Ag)	0	0	ne	ne
Boron, dissolved	6	6	1	<20	40	520	750 (Ag)	0	0	ne	ne
Cadmium, dissolved	4	4	4	<2	<2	<2	5 (MCL), 10 (Ag)	0	0	ne	ne
Copper, dissolved	3	3	0	16	27	<0.1	1,000 (SMCL), 200 (Ag)	0	0	ne	ne
Iron, dissolved	7	7	4	<10	<10	50	300 (SMCL), 5,000 (Ag)	0	0	ne	ne
Manganese, dissolved	1	1		<10	nc	110	50 (SMCL); 200 (Ag)	1; 0	1; 0	1978; ne	110; ne
Mercury, dissolved	7	7	7	<0.1	<0.1	<0.1	10 (Ag)	0	0	ne	ne
Molybdenum, dissolved	7	7	6	<1	nc	20	35 (HH)	0	0	ne	ne
Nickel, dissolved	1	1	1	<2	<2	<2	100 (HH), 200 (Ag)	0	0	ne	ne
Selenium, dissolved	7	7	6	<1	nc	6	50 (MCL), 20 (Ag)	0	0	ne	ne
Zinc, dissolved	6	6	0	20	270	830	5,000 (SMCL), 2,000 (Ag)	0	0	ne	ne

Table 24. Summary statistics for selected trace element water-quality data and exceedances of Colorado Department of Public Health and Environment water-quality standards for groundwater in the Upper Yampa River watershed, Colorado, 1975 through 1989 and 1998.—Continued

[No., number; Min., minimum; Max., maximum; <, less than; Ag, Agriculture; ne, no exceedance; nc, not computed; MCL, maximum contaminant level; SMCL, secondary maximum contaminant level; --, no water-quality standard; HH, human health. All constituents are reported in micrograms per liter. Number of significant figures for individual constituents may vary among the geologic units because data are from different sources and analytical periods. Water-quality standards are from Colorado Department of Public Health and Environment (2009b)]

Constituent	Total no. of wells	Total no. of samples	Total no. of censored values[1]	Min. value	Median value	Max. value	Water-quality standard	Exceedances of water-quality standard — No. of wells	No. of samples	Year(s)	Range of values not meeting standard
Curtis Formation of San Rafael Group											
Arsenic, dissolved	1	1	0	1	1	1	10 (MCL), 100 (Ag)	0	0	ne	ne
Boron, dissolved	1	1	0	30	30	30	750 (Ag)	0	0	ne	ne
Iron, dissolved	1	1	0	20	20	20	300 (SMCL), 5,000 (Ag)	0	0	ne	ne
Manganese, dissolved	1	1	0	20	20	20	50 (SMCL), 200 (Ag)	0	0	ne	ne
Mercury, dissolved	1	1	1	<0.1	<0.1	<0.1	10 (Ag)	0	0	ne	ne
Molybdenum, dissolved	1	1	0	5	5	5	35 (HH)	0	0	ne	ne
Selenium, dissolved	1	1	1	<1	<1	<1	50 (MCL), 20 (Ag)	0	0	ne	ne
Strontium, dissolved	1	1	0	560	560	560	--	--	--	--	--
Zinc, dissolved	1	1	1	<20	<20	<20	5,000 (SMCL), 2,000 (Ag)	0	0	ne	ne
Eocene series											
Arsenic, dissolved	1	1	1	<1	<1	<1	10 (MCL), 100 (Ag)	0	0	ne	ne
Cobalt, dissolved	1	1	1	<2	<2	<2	50 (Ag)	0	0	ne	ne
Iron, dissolved	1	1	0	80	80	80	300 (SMCL), 5,000 (Ag)	0	0	ne	ne
Manganese, dissolved	1	1	1	<10	<10	<10	50 (SMCL), 200 (Ag)	0	0	ne	ne
Mercury, dissolved	1	1	1	<0.5	<0.5	<0.5	10 (Ag)	0	0	ne	ne
Molybdenum, dissolved	1	1	1	<1	<1	<1	35 (HH)	0	0	ne	ne
Selenium, dissolved	1	1	1	<1	<1	<1	50 (MCL), 20 (Ag)	0	0	ne	ne
Fort Union Formation											
Arsenic, dissolved	1	1	0	1	1	1	10 (MCL), 100 (Ag)	0	0	ne	ne
Iron, dissolved	1	1	0	830	830	830	300 (SMCL), 5,000 (SMCL)	1	1	1975	830
Manganese, dissolved	1	1	0	40	40	40	50 (SMCL), 200 (Ag)	0	0	ne	ne
Selenium, dissolved	1	1	0	3	3	3	50 (MCL), 20 (Ag)	0	0	ne	ne
Lewis Shale											
Arsenic, dissolved	6	6	4	<1	<1	2	10 (MCL), 100 (Ag)	0	0	ne	ne
Boron, dissolved	1	2	0	190	285	380	750 (Ag)	0	0	ne	ne
Cadmium, dissolved	1	1	1	<1	<1	<1	5 (MCL), 10 (Ag)	0	0	ne	ne
Chromium, dissolved	1	1	0	10	10	10	100 (MCL, Ag)	0	0	ne	ne
Cobalt, dissolved	1	1	0	<2	<2	<2	50 (Ag)	0	0	ne	ne
Copper, dissolved	1	1	0	190	190	190	1,000 (SMCL), 200 (Ag)	0	0	ne	ne
Iron, dissolved	6	6	0	20	45	1,600	300 (SMCL), 5,000 (SMCL)	1	1	1975	1,600
Lead, dissolved	1	1	1	<2	<2	<2	50 (HH), 100 (Ag)	0	0	ne	ne
Manganese, dissolved	6	6	3	<10	15	310	50 (SMCL), 200 (Ag)	2 / 1	2 / 1	1975 / 1975	60, 310 / 310
Mercury, dissolved	6	6	5	<0.5	nc	1	10 (Ag)	0	0	ne	ne
Molybdenum, dissolved	5	5	3	<1	<1	7	35 (HH)	0	0	ne	ne
Selenium, dissolved	6	7	3	<1	1	13	50 (MCL), 20 (Ag)	0	0	ne	ne
Zinc, dissolved	6	7	0	20	50	1,700	5,000 (SMCL), 2,000 (Ag)	0	0	ne	ne

Table 24. Summary statistics for selected trace element water-quality data and exceedances of Colorado Department of Public Health and Environment water-quality standards for groundwater in the Upper Yampa River watershed, Colorado, 1975 through 1989 and 1998.—Continued

[No, number; Min, minimum; Max, maximum; <, less than; Ag, Agriculture; nc, not computed; MCL, maximum contaminant level; SMCL, secondary maximum contaminant level; --, no water-quality standard; HH, human health. All constituents are reported in micrograms per liter. Number of significant figures for individual constituents may vary among the geologic units because data are from different sources and analytical periods. Water-quality standards are from Colorado Department of Public Health and Environment (2009b)]

Constituent	Total no. of wells	Total no. of samples	Total no. of censored values[1]	Min. value	Median value	Max. value	Water-quality standard	Exceedances of water-quality standard			
								No. of wells	No. of samples	Year(s)	Range of values not meeting standard
Mancos Shale											
Arsenic, dissolved	12	13	9	<1	<1	2	10 (MCL), 100 (Ag)	0	0	ne	ne
Boron, dissolved	9	9	0	30	130	1,000	750 (Ag)	1	1	1978	1,000
Cadmium, dissolved	3	3	3	<2	<2	<2	5 (MCL), 10 (Ag)	0	0	ne	ne
Cobalt, dissolved	1	1	1	<2	<2	<2	50 (Ag)	0	0	ne	ne
Copper, dissolved	5	5	2	<2	22	100	1,000 (SMCL), 200 (Ag)	0	0	ne	ne
Iron, dissolved	12	13	2	<10	80	3,100	300 (SMCL)	4	4	1978	310–3,100
							5,000 (Ag)	0	0	ne	ne
Manganese, dissolved	12	13	6	<10	20	230	50 (SMCL)	3	3	1978	80–230
							200 (Ag)	1	1	1978	230
Mercury, dissolved	10	11	11	<0.1	nc	<0.5	10 (Ag)	0	0	ne	ne
Molybdenum, dissolved	10	11	5	<1	1	74	35 (HH)	1	1	1978	74
Nickel, dissolved	4	4	4	<2	<2	<2	100 (HH), 200 (Ag)	0	0	ne	ne
Selenium, dissolved	12	13	10	<1	<1	5	50 (MCL), 20 (Ag)	0	0	ne	ne
Strontium, dissolved	9	9	0	70	330	3,900	--	--	--	--	--
Zinc, dissolved	10	11	1	<20	220	4,500	5,000 (SMCL)	0	0	ne	ne
							2,000 (Ag)	2	3	1975, 1978	2,200–4,500
Mesaverde Group											
Aluminum, dissolved	46	58	11	<2	10	520	5,000 (Ag)	0	0	ne	ne
Antimony, dissolved	16	16	12	<1	<1	1	6 (MCL)	0	0	ne	ne
Arsenic, dissolved	39	82	26	<1	1	7	10 (MCL), 100 (Ag)	0	0	ne	ne
Boron, dissolved	33	119	0	30	130	1,000	750 (Ag)	2	3	1980, 1981	760–1,000
Cadmium, dissolved	22	45	42	<1	nc	13	5 (MCL), 10 (Ag)	3	3	1977	11–13
Cobalt, dissolved	1	1	1	<2	<2	<2	50 (Ag)	0	0	ne	ne
Copper, dissolved	21	24	11	<2	11	44	1,000 (SMCL), 200 (Ag)	0	0	ne	ne
Iron, dissolved	41	127	10	<3	50	39,000	300 (SMCL)	8	11	[3]1975–82	460–39,000
							5,000 (Ag)	1	2	1980, 1981	24,000, 39,000
Iron, total recoverable	46	53	0	70	1,400	590,000	--	--	--	--	--
Lead, dissolved	18	45	44	<1	nc	120	50 (HH), 100 (Ag)	1	1	1977	120
Manganese, dissolved	38	110	11	<1	60	3,000	50 (SMCL)	14	56	[3]1977–86	60–3,000
							200 (Ag)	8	39	[3]1977–86	220–3,000
Mercury, dissolved	39	81	71	<0.1	nc	2	10 (Ag)	0	0	ne	ne
Molybdenum, dissolved	33	69	46	<1	1	34	35 (HH)	0	0	ne	ne
Nickel, dissolved	11	12	12	<2	<2	<2	100 (HH), 200 (Ag)	0	0	ne	ne
Selenium, dissolved	41	128	104	<1	nc	7	50 (MCL), 20 (Ag)	0	0	ne	ne
Silver, dissolved	1	1	1	<1	<1	<1	50 (HH)	0	0	ne	ne
Strontium, dissolved	2	2	0	190	495	800	--	--	--	--	--
Zinc, dissolved	35	102	13	<3	20	2,200	5,000 (SMCL)	0	0	ne	ne
							2,000 (Ag)	1	1	1975	2,200

Table 24. Summary statistics for selected trace element water-quality data and exceedances of Colorado Department of Public Health and Environment water-quality standards for groundwater in the Upper Yampa River watershed, Colorado, 1975 through 1989 and 1998.—Continued

[No, number; Min., minimum; Max., maximum; <, less than; Ag, Agriculture; ne, no exceedance; nc, not computed; MCL, maximum contaminant level; SMCL, secondary maximum contaminant level; --, no water-quality standard; HH, human health. All constituents are reported in micrograms per liter. Number of significant figures for individual constituents may vary among the geologic units because data are from different sources and analytical periods. Water-quality standards are from Colorado Department of Public Health and Environment (2009b)]

Constituent	Total no. of wells	Total no. of samples	Total no. of censored values[1]	Min. value	Median value	Max. value	Water-quality standard	Exceedances of water-quality standard			
								No. of wells	No. of samples	Year(s)	Range of values not meeting standard
Precambrian erathem											
Arsenic, dissolved	1	1	1	<1	<1	<1	10 (MCL), 100 (Ag)	0	0	ne	ne
Copper, dissolved	1	1	1	<2	<2	<2	1,000 (SMCL), 200 (Ag)	0	0	ne	ne
Iron, dissolved	1	1	0	660	660	660	300 (SMCL), 5,000 (Ag)	1	1	1978	660
Manganese, dissolved	1	1	0	110	110	110	50 (SMCL), 200 (Ag)	1	1	1978	110
Mercury, dissolved	1	1	1	<0.1	<0.1	<0.1	10 (Ag)	0	0	ne	ne
Molybdenum, dissolved	1	1	1	<1	<1	<1	35 (HH)	0	0	ne	ne
Selenium, dissolved	1	1	1	<1	<1	<1	50 (MCL), 20 (Ag)	0	0	ne	ne
Strontium, dissolved	1	1	0	20	20	20	--	--	--	--	--
Zinc, dissolved	1	1	0	260	260	260	5,000 (SMCL), 2,000 (Ag)	0	0	ne	ne
Upper Cretaceous series											
Arsenic, dissolved	1	1	1	<1	<1	<1	10 (MCL), 100 (Ag)	0	0	ne	ne
Iron, dissolved	1	1	0	20	20	20	300 (SMCL), 5,000 (Ag)	0	0	ne	ne
Manganese, dissolved	1	1	0	60	60	60	50 (SMCL), 200 (Ag)	1	1	1975	60
Selenium, dissolved	1	1	0	3	3	3	50 (MCL), 20 (Ag)	0	0	ne	ne
Valley-fill deposits											
Arsenic, dissolved	6	6	6	<1	<1	<1	10 (MCL), 100 (Ag)	0	0	ne	ne
Copper, dissolved	1	1	0	300	300	300	1,000 (SMCL), 200 (Ag)	1	1	1975	300
Iron, dissolved	7	7	1	<10	140	3,100	300 (SMCL), 5,000 (Ag)	3	3	1975	1,300, 3,100
Lead, dissolved	1	1	1	<2	<2	<2	50 (HH), 100 (Ag)	0	0	ne	ne
Manganese, dissolved	7	7	3	<10	20	70	50 (SMCL), 200 (Ag)	1	1	1975	70
Mercury, dissolved	4	4	4	<0.5	<0.5	<0.5	10 (Ag)	0	0	ne	ne
Molybdenum, dissolved	4	4	3	<1	<1	2	35 (HH)	0	0	ne	ne
Selenium, dissolved	7	7	5	<1	<1	4	50 (MCL), 20 (Ag)	0	0	ne	ne
Zinc, dissolved	4	4	0	20	25	170	5,000 (SMCL), 2,000 (Ag)	0	0	ne	ne

Table 24. Summary statistics for selected trace element water-quality data and exceedances of Colorado Department of Public Health and Environment water-quality standards for groundwater in the Upper Yampa River watershed, Colorado, 1975 through 1989 and 1998.—Continued

[No, number; Min, minimum; Max, maximum; <, less than; Ag, Agriculture; ne, no exceedance; nc, not computed; MCL, maximum contaminant level; SMCL, secondary maximum contaminant level; --, no water-quality standard; HH, human health. All constituents are reported in micrograms per liter. Number of significant figures for individual constituents may vary among the geologic units because data are from different sources and analytical periods. Water-quality standards are from Colorado Department of Public Health and Environment (2009b)]

Constituent	Total no. of wells	Total no. of samples	Total no. of censored values[1]	Min. value	Median value	Max. value	Water-quality standard	Exceedances of water-quality standard			
								No. of wells	No. of samples	Year(s)	Range of values not meeting standard
						Unknown geologic units					
Aluminum, dissolved	103	268	155	<1	7	4,500	5,000 (Ag)	0	0	ne	ne
Antimony, dissolved	4	4	4	<1	<1	<1	6 (MCL)	0	0	ne	ne
Arsenic, dissolved	66	291	189	<1	<1	8	10 (MCL), 100 (Ag)	0	0	ne	ne
Barium, dissolved	103	215	0	14	69	680	2,000 (MCL)	0	0	ne	ne
Beryllium, dissolved	24	215	207	<0.5	nc	6	4 (MCL) / 100 (Ag)	1 / 0	2 / 0	1988, 1989	4.5, 6.0 / ne
Boron, dissolved	51	309	1	[2]10	150	1,400	750 (Ag)	1	1	1978	1,400
Cadmium, dissolved	45	255	213	<1	nc	23	5 (MCL) / 10 (Ag)	10 / 1	10 / 1	1988–89 / 1989	6–23 / 23
Chromium, dissolved	25	216	214	<5	nc	18	100 (MCL, Ag)	0	0	ne	ne
Cobalt, dissolved	103	218	203	<2	nc	23	50 (Ag)	0	0	ne	ne
Copper, dissolved	41	233	226	<2	nc	<50	1,000 (SMCL), 200 (Ag)	0	0	ne	ne
Iron, dissolved	62	329	18	<3	40	5,100	300 (SMCL) / 5,000 (Ag)	22 / 1	34 / 1	[3]1975–89 / 1981	340–5,100 / 5,100
Iron, total recoverable	20	51	0	280	4,000	190,000	--	--	--	--	--
Lead, dissolved	103	263	247	<1	nc	<50	50 (HH), 100 (Ag)	0	0	ne	ne
Manganese, dissolved	72	318	14	<1	60	2,600	50 (SMCL) / 200 (Ag)	42 / 21	174 / 84	[3]1975–89 / [3]1975–89	55–2,600 / 210–2,600
Mercury, dissolved	39	71	68	<0.1	nc	<0.5	10 (Ag)	0	0	ne	ne
Molybdenum, dissolved	60	283	255	<1	nc	<50	35 (HH)	0	0	ne	ne
Nickel, dissolved	31	223	211	<2	nc	<50	100 (HH), 200 (Ag)	0	0	ne	ne
Selenium, dissolved	71	332	293	<1	nc	11	50 (MCL), 20 (Ag)	0	0	ne	ne
Silver, dissolved	103	215	188	<1	nc	<30	50 (HH)	0	0	ne	ne
Strontium, dissolved	26	217	0	7	680	1,000	--	--	--	--	--
Zinc, dissolved	58	303	103	<3	9	1,200	5,000 (SMCL), 2,000 (Ag)	0	0	ne	ne

[1]Censored values can be expressed as values less than the laboratory reporting level.

[2]The minimum censored value is greater than the minimum detected value that is shown.

[3]Exceedances did not occur in every year of the period of record.

Table 25. Trace elements with Colorado Department of Public Health and Environment water-quality standards for groundwater and presence or absence of exceedances for samples collected in the Upper Yampa River watershed, Colorado, 1975 through 1989 and 1998.

[--, no water-quality standard. Number in parentheses after "Yes" is number of samples with exceedances of the water-quality standard]

Constituent	Exceedance of water-quality standard			
	Maximum contaminant level	Human-health	Secondary maximum contaminant level	Agricultural-use
Aluminum	--	--	--	No
Antimony	No	--	--	--
Arsenic	Yes (1)	--	--	Yes (1)
Barium	No	--	--	--
Beryllium	Yes (2)	--	--	No
Boron	--	--	--	Yes (70)
Cadmium	Yes (13)	--	--	Yes (4)
Chromium	No	--	--	No
Cobalt	--	--	--	No
Copper	--	--	No	Yes (1)
Iron	--	--	Yes (60)	Yes (3)
Lead	--	Yes (1)	--	Yes (1)
Manganese	--	--	Yes (284)	Yes (141)
Mercury	--	--	--	No
Molybdenum	--	Yes (1)	--	--
Nickel	--	No	--	No
Selenium	No	--	--	Yes (1)
Silver	--	No	--	--
Zinc	--	--	No	Yes (4)

dissolved iron SMCL were slightly more common in samples from wells in the Yampa coal field (28 percent of wells) than samples from wells not in the coal field (20 percent of wells). Only three samples, two from a well in the Mesaverde Group and one from a well in an unknown geologic unit, had concentrations that exceeded the dissolved iron agricultural-use standard of 5,000 µg/L.

Data for total recoverable iron were available for only 105 samples (table 24). A total of 51 and 53 samples were from unknown geologic units and the Mesaverde Group, respectively, and one sample was from the flood-plain alluvium. Median concentrations for samples from unknown geologic units and the Mesaverde Group were 4,000 and 1,400 µg/L, respectively. Concentrations greater than 10,000 µg/L were detected in samples from wells in unknown geologic units, the Mesaverde Group, and the flood-plain alluvium near Fish, Trout, and Grassy Creeks in the Yampa coal field. Reducing conditions in the subsurface in the coal field area could be causing the high concentrations of dissolved and total recoverable iron. The CDPHE has not established a total recoverable iron standard for groundwater.

Dissolved manganese data were available for all geologic units, especially unknown geologic units (318 samples), Mesaverde Group (110 samples), and terrace alluvium (67 samples) (table 24). Median concentrations were 30 µg/L or less for samples from the flood-plain alluvium, Browns Park

Formation, and Lewis and Mancos Shales; the median for the Browns Park Formation was the lowest at less than 10 µg/L. Median dissolved manganese concentrations for the terrace alluvium (70 µg/L), Mesaverde Group, and unknown geologic units (60 µg/L each) were greater than the CDPHE SMCL for manganese of 50 µg/L, as were single concentrations in samples from the Upper Cretaceous series (60 µg/L), valley-fill deposits (70 µg/L), and Precambrian erathem (110 µg/L) (table 25). In total, the CDPHE SMCL for dissolved manganese was exceeded in concentrations from more than one-half (284 of 548) of the samples. Between 51 and 57 percent of samples each from the terrace alluvium, Mesaverde Group and unknown geologic units had concentrations greater than the SMCL (table 24). Samples from most (86 percent) of the wells in the terrace alluvium had concentrations greater than the standard. Exceedances also were common for wells in the Mesaverde Group (37 percent of wells) and unknown geologic units (58 percent of wells). Wells with concentrations exceeding the SMCL were concentrated in the Yampa coal field (58 percent of wells) and not in noncoal field areas (26 percent of wells). About 26 percent of groundwater samples also had dissolved manganese concentrations that exceeded the CDPHE agricultural-use standard (table 24).

Analysis of concentration data for groundwater samples indicates that water from wells in the flood-plain alluvium and Browns Park Formation would be more suitable for domestic or agricultural use than water from wells in the Lewis and Mancos Shales and Mesaverde Group. Suitability of water from wells in the Curtis Formation, Eocene series, Fort Union Formation, Precambrian erathem, Upper Cretaceous series, and valley-fill deposits for domestic or agricultural use could not be determined because too few data were available for analysis.

Macroinvertebrate Data

Data for macroinvertebrate communities and population were available for 66 stream sites (62 unique site locations) in the UYRW for various periods of time between 1975 and 2008 (fig. 18, appendix 6). Data consist of counts of the number of individuals within a given taxa. About 38 percent (25 of 66) of sites were sampled once by the USGS during August and September 1975. These data are available at *http://rmgsc.cr.usgs.gov/cwqdr/Yampa/index.shtml*, accessed June 2012). The CDPHE collected macroinvertebrate data at 37 sites during most years from 1997 through 2008. Sample collection occurred during April and July through October, and three or fewer samples were collected at each site. Data were made available to this study from the state of Colorado Ecological Data Application System (Chris Theel, Colorado Department of Public Health and Environment, written commun., 2009). Most of the data were included in an Ecological Monitoring and Assessment Program report (Beyea and Theel, 2007). Data on macroinvertebrate communities and population have also been collected by GEI Consultants, Inc., for the City of Steamboat

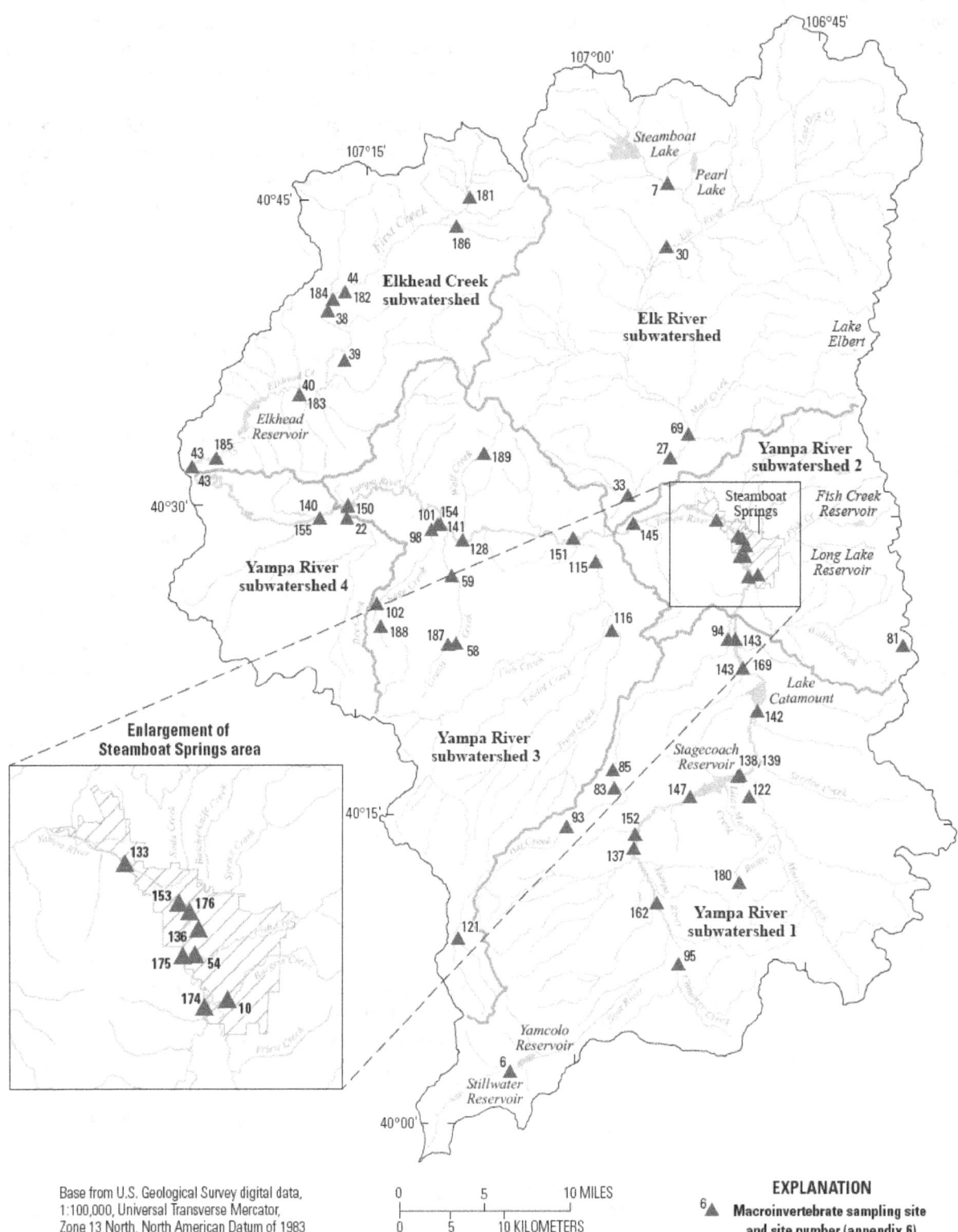

Figure 18. Location of stream sites with macroinvertebrate data, Upper Yampa River watershed, Colorado, 1975 through 2008.

Springs. Population data for various taxa were available for four sites on the Yampa River within the city limits for one day each during the middle of September 2005 and 2007, and late August 2008 (GEI Consultants, Inc., 2007, 2008). Data were collected to determine whether changes in the macroinvertebrate community occurred throughout the study reach or over time. Macroinvertebrate data from the CDPHE and GEI Consultants, Inc., are available at the Colorado Data Sharing Network website *http://www.coloradowaterdata.org/ awqmsportal.html* (accessed June 2012), which is a project of the Colorado Water Quality Monitoring Council.

Detailed analysis of the macroinvertebrate data was beyond the scope of this report. However, results of macroinvertebrate sampling at four Yampa River sites within the Steamboat Springs city limits from the 2008 GEI Consultants, Inc., report are summarized in this paragraph (GEI Consultants, Inc., 2008). For the 2008 sampling effort, density (number of individuals per square meter) was the highest at the most downstream site. Insects were the most commonly collected taxa throughout the study reach. Water mites, crustaceans, snails, and clams occurred in small numbers in the reach. Segmented worms (Family Naididae) increased in density from upstream to downstream and represented, on average, about 43 percent of the individuals collected at the most downstream site. Naid worms are more tolerant of organic pollution than beetles and caddisflies, the dominant taxon at the upstream sites (GEI Consultants, Inc., 2008). Most of the macroinvertebrates collected in the study reach, however, were tolerant of diverse environmental conditions. Many of the changes in community characteristics during the years from 2005 through 2008 occurred throughout the study reach. This may indicate that changes in community characteristics throughout the reach are due to upstream effects or large-scale environmental factors rather than changes in water quality within the reach (GEI Consultants, Inc., 2008).

Synthesis of Water-Quality Data in the Upper Yampa River Watershed

The assessment of water quality of streams, lakes, reservoirs, and groundwater in the UYRW was based on data collected at selected sites for various periods from 1975 through 2009. Data were retrieved from the UYRW water-quality database (*http://rmgsc.cr.usgs.gov/cwqdr/Yampa/index.shtml*, accessed June 2012). A number of quality-assurance procedures were applied to the data. For a large portion of the water-quality data, however, limited metadata and (or) quality-assurance data were available. For these data, it is assumed that measurements made at the time of sampling and results from laboratory analyses are of good quality.

The number of sites sampled in the UYRW and the amount of data collected varied spatially and temporally. Results of the assessment indicate that the quality of surface water is likely controlled by three primary factors: geology,

streamflow, and land use. The quality of groundwater is a function of various physical and geochemical processes, including precipitation, the depositional environment of aquifer sediment, type of sediments that groundwater moves through, dissolution of soluble minerals in rocks and soils, and ion exchange reactions.

The water quality of surface water in the UYRW depends in large part on the geology underlying the drainage basin for a stream or water body, which, in turn, affects the quality of groundwater discharging to surface water. Drainage basins in the eastern one–third of the UYRW, most commonly those in Yampa River subwatershed 2 and the Elk River subwatershed, are underlain primarily by igneous and metamorphic rocks that are resistant to the weathering action of water. With less weathering, a small amount of dissolved constituents may be present in surface water and groundwater. Median values of specific conductance, hardness, ANC, dissolved solids, and suspended sediment for stream samples from Yampa River subwatershed 2 and the Elk River subwatershed were lowest for all subwatersheds in the UYRW. Water samples from high-altitude lakes and reservoirs in these two subwatersheds also had very low values for specific conductance (maximum value of 34 µS/cm) and hardness and ANC (maximum values of 17.9 mg/L or less). Low ANC values for streams and water bodies in the eastern one-third of the watershed and other areas such as the Flat Tops Wilderness area indicate that some surface waters in the UYRW are sensitive to inputs of acidic deposition from precipitation, primarily because the water, soil types, and rocks have little capacity to buffer acidic inputs. Groundwater samples from two wells in Yampa River subwatershed 2 that tap in the Precambrian erathem, a geologic unit composed of igneous and metamorphic rocks, had the lowest median specific conductance (163 µS/cm) for the sampled geologic units in the UYRW.

Drainage basins in the western two-thirds of the watershed, including Yampa River subwatersheds 3 and 4 and the Elkhead Creek subwatersheds, are underlain primarily by Cretaceous- and Tertiary-age sedimentary rocks (primarily sandstones and shales) deposited in marine and nonmarine environments. The sedimentary rocks are more susceptible to the weathering action of water than igneous and metamorphic rocks, which can result in a large amount of dissolved constituents in surface water and groundwater in areas with sedimentary rocks. Materials weathered from sedimentary rocks also can contain a large amount of trace elements. Median values of specific conductance, hardness, ANC, dissolved solids, and suspended sediment for stream samples from Yampa River subwatersheds 3 and 4 were highest for all subwatersheds in the UYRW. Median values of specific conductance for samples from wells completed in the sedimentary Lewis Shale and the Mesaverde Group, primarily in Yampa River subwatersheds 3 and 4, were 1,000 µS/cm or more. These high values can occur in surface water and groundwater because of natural conditions.

Differences in specific conductance and median concentrations of dissolved constituents (dissolved solids, sulfate, chloride, and others) in groundwater from the sedimentary geologic units with more than one sample reflect differences in the depositional environment of aquifer sediments, type of sediments that groundwater moves through, dissolution of soluble minerals in rocks and soils, and ion exchange reactions. Specific conductance was typically lower in samples from wells in the Browns Park Formation than in samples from wells in other sedimentary geologic units, as were median concentrations of dissolved solids, and dissolved sulfate and chloride. This most likely is due to the riverine depositional environment of the Browns Park Formation, rather than a marine or marine-nonmarine environment, and the absence of shale. The highest median concentrations of dissolved sulfate and chloride were in samples from wells completed in the terrace alluvium. These wells are located in one small area of the UYRW downstream from Sage Creek. Elevated sulfate and chloride concentrations could result from insufficient leaching of material from the alluvium because of the semi-arid climate. Elevated sulfate concentrations could also occur naturally because of geochemical processes involving gypsum and oxidation of other minerals in the subsurface.

Values of pH in stream water and groundwater samples less than the CDPHE standard minima were not as common in the UYRW as in other mountain watersheds in Colorado. Unlike the UYRW, low pH values in stream water and groundwater in other mountain areas can occur naturally from mineralized bedrock and (or) result from acidic drainage in hard-rock mining areas (Runkel and others, 2007). Mineralized bedrock and hard-rock mining are not common in the UYRW. In the Yampa coal field, primarily in Yampa River subwatersheds 3 and 4, all pH values in stream-water samples with the exception of one were greater than 7.1, and about 96 percent of well samples had pH values greater than 6.5. The absence of acidic coal-mine drainage in the UYRW when compared to other areas in the country is likely because of the low sulfur content of Colorado coal (Wentz, 1974). With a low amount of sulfur, sulfuric acid is unlikely to form in streams and groundwater. Stream water and groundwater in the Yampa coal field also have a high capacity to buffer inputs of acidic water.

Attainment of CDPHE water-quality standards for surface water was met for many stream water sites. Attainment of the aquatic-life standard was not met at 15 sites for dissolved copper, total recoverable iron, or dissolved selenium. Only one or two groundwater samples had concentrations of dissolved arsenic, lead, molybdenum, and beryllium that exceeded CDPHE standards, and concentrations of dissolved cadmium exceeded the standard in less than 4 percent of samples. Exceedances of standards for dissolved iron and manganese were more common. Attainment of CDPHE stream water-quality standard for these two trace elements was not met at 3 and 18 sites, respectively. The CDPHE SMCLs for dissolved iron and manganese in groundwater were exceeded

in about 10 and more than 50 percent of samples, respectively. Elevated concentrations of dissolved iron and manganese can be common in surface water and groundwater. Iron is associated with a variety of minerals in igneous and sedimentary rocks and is present in organic materials. Many of the minerals with iron also have manganese. Iron and manganese in groundwater are probably affected by reduction-oxidation processes. Portions of five stream segments in the UYRW are on the CDPHE 2012 303(d) lists for impairment or monitoring or evaluation for trace element contamination, specifically for total recoverable iron, lead, manganese, mercury, selenium, and (or) zinc.

Changes in streamflow drive seasonal variability in specific conductance and other properties and constituents. Values of specific conductance, for example, were lowest during snowmelt runoff when there was increased dilution and little interaction between water and soil or rock and higher during most other times of the year when streamflow is lower and base flow from groundwater is the major source of water in a stream. Concentrations of constituents that bind to particulate matter, such as phosphorus and iron, can be higher during the initial phase of snowmelt runoff as material that is washed off the land surface drains into streams. Similarly, concentrations of suspended sediment were highest during snowmelt runoff.

In the UYRW, as in other mountainous watersheds in Colorado, land and water use and changes in the uses can also affect water quality of a stream, including water temperature and the presence of total phosphorus and *E. coli*. One issue of concern related to these changes is water temperature in the main stem of the Yampa River. Many temperature values measured at the Yampa River at Steamboat Springs (site 153) from 2002 through 2005, particularly those measured during late July and early August, were not in attainment of the CDPHE June–September or October–May standards for cold water. Higher water temperature could be the result of reduced streamflow during drought but also could result from reduced streamflow due to upstream hydrologic modifications and changes in the river channel. The Yampa River from the confluence with Oak Creek downstream to the confluence with Elkhead Creek is on the CDPHE 2012 303(d) monitoring and evaluation list due to high water temperature, which can stress aquatic life. Aquatic life can also be affected by large amounts of nutrients that promote excessive algal growth (eutrophication) and deplete oxygen levels in a water body.

Concentrations of unfiltered total phosphorus greater than laboratory detections levels were detected in samples collected from streams throughout the watershed. Sources of phosphorus to streams in the UYRW can include some sedimentary rocks (coal beds in the Mesaverde Group), urban and agricultural runoff, wastewater-treatment-plant effluent, and animal waste. About 14 percent of samples had individual values of unfiltered total phosphorus that were greater than USEPA recommendations to prevent eutrophication. These samples were collected in Yampa River subwatersheds 1, 2, and 3 and the

Elkhead Creek subwatershed. There was a statistically significant upward trend in flow-adjusted concentrations of unfiltered total phosphorus for Yampa River at Steamboat Springs (site 153) for 1997 through 2008. From 2000 through 2008, the population of Routt County grew by more than 18 percent and was largely driven by recreation-related tourism, including second-home development and conversion of undeveloped or agricultural land to urban land.

The bacterium *E. coli*, which can come from recreational water users, wildlife, livestock, and septic systems, was detected at concentrations greater than the CDPHE water-quality standard in five samples collected from three sites during 1994, 1999, 2001, and 2003. Three stream segments are on the CDPHE 2012 303(d) list of impaired waters or the monitoring and evaluation list for *E. coli*.

Many of the physical properties or constituents analyzed for this study probably would not be issues of concern for aquatic life, human health, or agricultural use of surface water and groundwater; concentrations at many sites were less than laboratory detections levels or less than CDPHE water-quality standards. Exceedances of standards for some constituents also probably would not be issues of concern because exceedances were based on older data (data collected before 20 years ago), although recent data are not available for verification. For example, data for four of five sites not in attainment of the CDPHE total recoverable iron standard for streams were collected before 1983. Periods of exceedances are listed in tables 10 and 21–24. Exceedances for some constituents, arsenic in groundwater, for example, may occur for only one or a few samples and may represent very localized water-quality conditions. Constituents that are issues of concern for aquatic life, human health, or suitability of water for various uses include those on the CDPHE 2012 303(d) list of impaired waters or monitoring or evaluation list for surface water. Other constituents in stream water or groundwater that are or could be issues of concern include those that commonly have concentrations that exceed standards or that could affect technical qualities of water. This could include unfiltered sulfate, unfiltered total phosphorus, and dissolved copper in stream water; pH, unfiltered nitrate plus nitrite, and dissolved copper in groundwater; and hardness, dissolved solids, iron, and manganese in stream water and groundwater.

This assessment of water-quality data for streams and groundwater in the UYRW illustrates the need for consistent (throughout a year and for many years) long-term data collection at key sites for temporal trend analysis and to identify emerging changes in water quality. Only three stream sites in the UYRW had sufficient (10 years or more of at least quarterly data collection ending after 2000 and less than 10 percent censored data) long-term data to test for temporal trends in specific conductance. Trend analysis of data for other constituents, including other physical properties, dissolved solids, selected major ions, total phosphorus, and some trace elements, could be conducted for only one site because of data requirements. Few stream sites had data for 10 years or more.

For many sites, there were yearly gaps in data collection. Analysis of water-quality data for groundwater was restricted to data collected from 1975 through 1989 and during 1998. No groundwater-quality data were in the UYRW water-quality database for 1990 through 1997, and there were no recent data (collected after 1998) in the database. About 66 percent of wells were sampled only once. Additional information on dissolved oxygen concentrations in groundwater and geologic units and aquifers that wells tap would be helpful for assessment of groundwater quality. Data on dissolved oxygen concentrations are needed for determining reduction-oxidation processes in groundwater; the concentration data only were available for 10 percent of wells sampled. Comparison of groundwater-quality data to geologic units and aquifers that wells tap was restricted to about 70 percent of wells. Well-completion information of this type was not available for about 30 percent of wells with water-quality data.

Summary

The Yampa River is a highly valued resource known for its biological diversity, largely unaltered natural condition, and generally high water quality. The Upper Yampa River watershed (UYRW, the Elkhead Creek subwatershed and the Yampa River watershed upstream from Elkhead Creek) is undergoing increased land and water development to support growing municipal demands, recreational tourism, and second-home development that present water-quality challenges. In response to the needs of stakeholders in the UYRW, the U.S. Geological Survey, in cooperation with Routt County, the Colorado Water Conservation Board, and the City of Steamboat Springs, initiated a study in 2009 to compile water-quality and macroinvertebrate data for the UYRW and assess water-quality conditions. For selected physical properties and chemical constituents in samples collected from streams, lakes, reservoirs, and groundwater wells in the UYRW, this report: (1) characterizes available data through statistical summaries, (2) evaluates the spatial and temporal distribution of water-quality conditions, (3) identifies temporal trends in water quality, where possible, (4) provides comparisons of the data to federal and state water-quality standards and recommendations, and (5) identifies factors affecting the quality of water. In addition, the availability and characteristics of macroinvertebrate data collected in the UYRW are described.

The UYRW drains approximately 1,800 square miles west of the Continental Divide in northwestern Colorado. The watershed extends from the Williams Fork and Flat Top Mountains to the Gore and Park Ranges and from the Continental Divide to, and including, the Elkhead Creek subwatershed. The UYRW is almost entirely contained within Routt County, with small portions in Grand, Garfield, Jackson, Moffat, and Rio Blanco Counties.

Water-quality data for selected physical properties and chemical constituents in samples collected by federal, state, and local agencies from streams, lakes, reservoirs, and ground-water wells in the UYRW for various periods from 1975 through 2009 are assessed. The data used in the assessment are available in a web-based repository at *http://rmgsc.cr.usgs.gov/ cwqdr/Yampa/index.shtml*. A number of quality-assurance procedures were applied to the data. For a large portion of the water-quality data, however, limited metadata and (or) quality-assurance data were available. For all data that could not be quality assured, it is assumed that measurements made at the time of sampling and results from laboratory analyses are of good quality.

Stream water-quality data collected by seven federal, state, or local agencies from 1975 through 2009 were compiled for 211 sites (176 unique site locations). Because streams in the UYRW can have distinct water-quality characteristics based on location in the watershed, the watershed was divided into six subwatersheds for data analysis:

- Yampa River and tributaries upstream from Chuck Lewis State Wildlife Area (Yampa River subwatershed 1);

- Yampa River and tributaries from Chuck Lewis State Wildlife Area to Elk River confluence (Yampa River subwatershed 2);

- Elk River and tributaries (Elk River subwatershed);

- Yampa River and tributaries from Elk River confluence to Town of Hayden (Yampa River subwatershed 3);

- Yampa River and tributaries from Town of Hayden to Elkhead Creek confluence (Yampa River subwatershed 4); and

- Elkhead Creek and tributaries (Elkhead Creek subwatershed).

A total of 5,861 stream-water samples were included in the assessment; data were collected from 1975 through 2009 for physical properties, dissolved solids, major ions, nutrients, trace elements, uranium, coliform bacteria, and suspended sediment. Data were collected every year at only one main-stem site on the Yampa River. Physical-property data were available for almost all samples. Samples for dissolved solids, major ions, nutrients, and trace elements also were commonly collected. About 13 percent of sites had data for more than 50 samples; in contrast, almost one-half of sites had data for five or fewer samples.

Physical properties in stream water analyzed for this study included specific conductance, pH, water temperature, dissolved oxygen, hardness, and acid neutralizing capacity. Specific conductance, which is proportional to the concentration of major dissolved constituents in water, was commonly low (less than 200 microsiemens per centimeter at 25 degrees Celsius (μS/cm)) in headwater tributaries with drainage basins that overlie igneous and metamorphic rocks that are resistant to the weathering action of water. Values greater than

1,000 μS/cm were most common in areas with sedimentary rocks that are susceptible to weathering. There was a statistically significant downward trend in specific conductance at the site 153 (Yampa River at Steamboat Springs).

The pH of a water sample is a measure of how acidic or alkaline the water is. Median pH values were lowest in the Elk River subwatershed and highest in Yampa River subwatershed 1 and the Elkhead Creek subwatershed. Fewer than 3 percent of pH values were outside the acceptable range established by the Colorado Department of Public Health and Environment (CDPHE) for protection of aquatic life. Water temperature varied depending on the altitude of a site and climate at the site. Water temperature for main-stem Yampa River sites increased in a downstream direction. For Yampa River at Steamboat Springs (site 153), water temperature on some days did not meet CDPHE acute and chronic cold-water standards for aquatic-life protection for the periods of June through September and October through May. This was the only site in the UYRW with a continuous record of water temperature. Dissolved oxygen is the measurement of the oxygen in water that is available to aquatic life. Concentrations of dissolved oxygen in UYRW streams were inversely related to water temperature. About 99 percent of sites met the CDPHE aquatic-life standard for dissolved oxygen. Streams with drainage basins underlain by igneous and metamorphic rocks tended to have softer water and a lower capacity to neutralize inputs of acidic water than streams with drainage basins underlain by sedimentary rocks.

The spatial distribution of dissolved solids and major ions (calcium, magnesium, sodium, potassium, bicarbonate, carbonate, sulfate, chloride, fluoride, and silica) in UYRW stream water was similar to that for specific conductance. All but about 4 percent of sites were in attainment of the CDPHE water-supply standard for unfiltered sulfate. Concentrations of major ions and dissolved solids in streams were typically lower during snowmelt runoff in May and June than at other times of the year primarily because of the water source and increased volume of streamflow. There was a statistically significant downward trend in dissolved solids concentrations at Yampa River at Steamboat Springs (site 153).

Nitrogen and phosphorus are essential nutrients for plants and animals, but high concentrations can cause excessive algal growth in surface waters. About 85 percent of dissolved nitrite concentrations in streams throughout the watershed were 0.01 milligram per liter (mg/L) or less, and attainment of the CDPHE standard for aquatic-life protection was met for all sites. More than 50 percent of dissolved nitrate concentrations were reported as less than laboratory detection levels (censored). Median concentrations were highest in Yampa River subwatersheds 3 and 4. All sites with data for unfiltered nitrate were in attainment of CDPHE standards for drinking water or agricultural use. More than 60 percent of unfiltered total ammonia concentrations in stream samples were less than detection levels. Less than one percent of samples had concentrations that exceeded the CDPHE aquatic-life standard; these samples were collected during 1975 or 1976. Median concentrations of unfiltered total phosphorus were highest in Yampa River subwatersheds 1, 3, and 4. Total phosphorus concentrations greater than federal recommendations to control downstream eutrophication were detected in about 14 percent of samples collected from more than 29 sites. A statistically significant upward trend in unfiltered total phosphorus concentrations was identified for Yampa River at Steamboat Springs (site 153). This trend may reflect population growth and land-use changes that have occurred upstream from the site.

Many trace elements are essential nutrients required by biota in small amounts, but substantial concentrations of trace elements can be toxic to aquatic life and possibly to wildlife, livestock, and people. For UYRW stream sites, about two-thirds of the concentration data for dissolved and total recoverable cadmium, lead, nickel, and silver; dissolved chromium, copper, selenium, and zinc; and total dissolved and recoverable mercury were less than detection levels. Concentrations greater than 1,000 micrograms per liter (μg/L) were detected for total recoverable aluminum, iron, manganese, and zinc, and dissolved iron, manganese, and strontium. Maximum concentrations of the various trace elements occurred in Yampa River subwatersheds 1, 2, 3, or 4 and seemed to depend on the lithology of the rocks underlying the subwatershed. Seasonal variation in concentrations was evident for total recoverable aluminum, copper, iron, and zinc. There was a statistically significant downward trend in total recoverable manganese at site 153. Some sites were not in attainment of CDPHE aquatic-life or water-supply standards for dissolved copper, iron, manganese, and selenium and for total recoverable iron.

Concentrations of the bacterium *E. coli* in five stream samples collected from 1994 through 2003 exceeded the CDPHE recreation standard of 126 colonies per 100 milliliters. Exceedances could be due to recreational users of a stream, wildlife, and (or) livestock.

The amount and size of suspended sediment in water are affected by streamflow. In the UYRW, suspended-sediment concentrations typically were lowest from August through February when streamflow was lowest and higher during April or May. The highest concentrations were in streams with watersheds underlain by sedimentary rocks.

Water-quality data for Lake Elbert, Long Lake Reservoir, Stagecoach Reservoir, Steamboat Lake, and Elkhead Reservoir for various time periods were analyzed or summarized for this study. Water in Lake Elbert and Long Lake Reservoir was very dilute; all values for specific conductance from 1985 through 2005 (Long Lake Reservoir) or through 2009 (Lake Elbert) were 34 μS/cm or less. The reservoirs have little capacity to neutralize inputs of acidic water. Exceedances of CDPHE standards for pH and dissolved iron and manganese were rare and only occurred on as many as 5 sample days for Long Lake Reservoir. Stagecoach Reservoir and Steamboat Lake were vertically stratified during July 2006. Dissolved oxygen concentrations less than 0.5 mg/L at depth indicate anoxic conditions. Measurements of dissolved oxygen in Elkhead Reservoir from July 1995 through August 2001 indicate anoxic conditions were present in the reservoir at times during stratification. Trophic status of the reservoir ranged from oligotrophic to eutrophic; phosphorus was the limiting nutrient in more samples (52 percent) than nitrogen (9 percent).

A total of 816 groundwater-quality samples collected from 328 wells during 1975 through 1989 and 1998 were analyzed for this study. The sampled wells are concentrated in the middle latitudes of the watershed. About 66 percent of the wells with water-quality data were sampled only once. Samples were collected most often during 1975, 1978, and 1988. Analysis of groundwater data focuses on one sample per day for physical properties, dissolved solids, major ions, nutrients, and trace elements. Data indicate that these wells tap aquifers in 12 geologic units, most often the flood-plain alluvium and Mesaverde Group. Well-construction information was not available for about 30 percent of wells. Wells without this information are grouped for this study as tapping "unknown geologic units." More groundwater samples were collected from wells completed in the unknown geologic units, Mesaverde Group, and terrace alluvium than from wells completed in the other geologic units.

Analysis of physical-property data for groundwater indicates that specific conductance was lower in samples collected from igneous and metamorphic rocks and sedimentary rocks of nonmarine origin than sedimentary rocks of marine or marine-nonmarine origin. Values of pH less than the lower bound of the CDPHE secondary maximum contaminant level (SMCL) for groundwater were not as common as values greater than the upper bound. Samples with pH values not meeting the standard were most commonly collected from the flood-plain alluvium (values less than 6.5) and unknown geologic units (values greater than 8.5). The acid neutralizing capacity of water in samples from the sedimentary rock units indicate that water in these rock units is well buffered.

Median dissolved solids concentrations were lowest for samples collected from valley-fill deposits and highest for samples collected from the Mesaverde Group. Dominant cations were calcium and (or) calcium plus magnesium of a mixture of these with sodium. Bicarbonate was the dominant anion, but many samples also had sulfate. Median dissolved sulfate concentrations were lowest in samples collected from the Browns Park Formation and highest in samples collected from the terrace alluvium. About one-half of the dissolved sulfate samples had concentrations that were greater than the CDPHE SMCL. Exceedances were most common for samples collected from wells in the terrace alluvium, Mesaverde Group, and unknown geologic units.

All dissolved nitrite concentrations in groundwater samples were well below CDPHE maximum contaminant level (MCL). Median concentrations of dissolved nitrate pus nitrite for all geologic units were 1.5 mg/L or less. About 4.5 percent of samples had concentrations greater than the CDPHE MCL for nitrate. Almost all dissolved and unfiltered total phosphorus concentrations were less than 0.1 mg/L.

Concentrations of many trace elements in groundwater samples were low; more than 80 percent of the samples collected for dissolved antimony, arsenic, beryllium, cadmium, chromium, cobalt, copper, lead, mercury, molybdenum, nickel, selenium, and silver had concentrations less than or equal to laboratory detection levels. All concentrations of dissolved antimony, barium, chromium, cobalt, mercury, nickel, and silver met CDPHE human-health (HH) standards, MCLs, SMCL, and (or) agricultural-use standards for groundwater. Only one or two groundwater samples each had concentrations of dissolved arsenic, beryllium, copper, lead, molybdenum, and selenium that exceeded CDPHE HH standards, MCLs, SMCL, and (or) agricultural-use standards. Concentrations of dissolved cadmium greater than the CDPHE MCL were detected in less than 4 percent of samples. All samples but one were from wells in the Yampa coal field.

The highest trace element concentrations (3,600 µg/L or more) in groundwater were detected for dissolved aluminum and boron, dissolved and total recoverable iron, and dissolved manganese, strontium, and zinc. All dissolved aluminum concentrations were less than the CDPHE agricultural-use standard for livestock watering. The agricultural-use standard for dissolved boron was exceeded in about 13 percent of samples, most commonly for the terrace alluvium. All zinc concentrations were less than the CDPHE SMCL; fewer than one percent of samples had zinc concentrations that exceeded the agricultural-use standard. Standards have not been established for dissolved strontium and total recoverable iron in groundwater.

Median concentrations of dissolved iron in geologic units with more than one sample were lowest for samples collected from the Browns Park Formation and highest for samples collected from valley-fill deposits. Concentrations of dissolved iron exceeded the CDPHE SMCL in about 10 percent of samples, most commonly those collected from valley-fill deposits and the Mancos Shale. Only three samples, two collected from the Mesaverde Group and one collected from an unknown geologic unit, had a dissolved-iron concentration that exceeded the CDPHE agricultural-use standard. Elevated concentrations (greater than 10,000 µg/L) of total recoverable iron were detected in samples from wells located in the Yampa coal field. Reducing conditions in the subsurface in the coal field area could be causing the high concentrations of iron.

The most exceedances of a CDPHE standard for trace elements in groundwater were for dissolved manganese; more than one-half of the samples had concentrations that exceeded the SMCL for groundwater. These samples were most commonly collected from the terrace alluvium, Mesaverde Group, and unknown geologic units. More than one-half of the exceedances were for wells in the Yampa coal field. About 26 percent of groundwater samples also had dissolved manganese concentrations that exceeded the CDPHE agricultural-use standard.

Macroinvertebrate community and population data were available for 66 stream sites in the UYRW for various periods of time between 1975 and 2008. A summary of results from one study of Yampa River sites in Steamboat Springs indicates that species tolerant of organic pollutants were more common at downstream sites than at upstream sites. However, changes observed in community characteristics between 2005 and 2008 may be due to upstream effects or large-scale environmental factors rather than changes in water quality within the stream reach.

Synthesis of water-quality data indicates that the values and concentrations of many physical properties and constituents in surface-water samples for the UYRW are likely controlled primarily by geology, streamflow, and land use. The quality of groundwater in the UYRW is a function of various physical and geochemical processes, including precipitation, the depositional environment of the aquifer sediments, type of sediments that groundwater moves through, dissolution of soluble minerals in rocks and soils, and ion-exchange reactions. Constituents that are issues of concern for aquatic life, human health, or suitability of water for various uses include those on the CDPHE 2012 303(d) list of impaired waters or monitoring or evaluation list for surface water. Other constituents in stream water or groundwater that are or could be issues of concern include those that commonly have concentrations that exceed standards or that could affect technical qualities of water. This could include unfiltered sulfate, unfiltered total phosphorus, and dissolved copper in stream water; pH, unfiltered nitrate plus nitrite, and dissolved copper in groundwater; and hardness, dissolved solids, iron, and manganese in stream water and groundwater. Analysis of stream-water and groundwater data for changes in water quality over time was limited because of the absence of long-term data collection in the UYRW. Consistent long-term data collection of many years is needed at key sites for temporal trend analysis of water-quality data and to identify emerging changes in water quality.

Acknowledgments

The authors thank Judith Thomas for her assistance with compilation of water-quality data into the UYRW water-quality database, and Barbara C. Ruddy for her work in preparing the land cover and geology maps for this report. Technical assistance was provided by Kenneth J. Leib and David K. Mueller. Technical reviews were provided by Peter B. McMahon, Allan D. Druliner, and Lisa D. Miller. Editorial review was provided by Ruth Larkin. Manuscript preparation was provided by Joy Monson, Loretta Ulibarri, and Carol Wilkinson. All of the personnel identified above are USGS employees.

References Cited

Adriano, D.C, 2001, Trace elements in terrestrial environments—Biogeochemistry, bioavailability, and risks of metals: New York, Springer–Verlag, 871 p.

Affolter, R.H., 2000, Quality characterization of Cretaceous coal from the Colorado Plateau coal assessment area, *in* Kitschbaum, M.A., Roberts, L.N.R., and Biewick, L.R.H., eds., Geologic assessment of coal in the Colorado Plateau—Arizona, Colorado, New Mexico, and Utah: U.S. Geological Survey Professional Paper 1625–B, p. G1–G136, accessed September 2010, at *http://pubs.usgs.gov/pp/p1625b/Reports/Chapters/Chapter_G.pdf.*

Bass, N.W., Eby, J.B., and Campbell, M.R., 1955, Geology and mineral fuels of parts of Routt and Moffat Counties, Colorado: U.S. Geological Survey Bulletin 1027–D, p. 143–250.

Beyea, B.W., and Theel, Christopher, 2007, Ecological monitoring and assessment report: Colorado Department of Public Health and Environment, Water Quality Control Division, 46 p., accessed June 2012, at *http://cospl.coal-liance.org/fez/eserv/co:1862/he17202ec72007internet.pdf.*

Blinn, D.W., and Poff, N.L., 2005, Yampa River, *in* Benke, A.C., and Cushing, C.E., eds., Rivers of North America: Elsevier Academic Press, p. 503–506.

Brodgen, R.E., and Giles, T.F., 1977, Reconnaissance of ground-water resources in a part of the Yampa River basin between Craig and Steamboat Springs, Moffat, and Routt Counties, Colorado: U.S. Geological Survey Water-Resources Investigations 77–4 map, scale 1:120,000.

Brownfield, M.E., Johnson, E.A., Affolter, R.H., and Barker, C.E., 1999, Coal mining in the 21st century, Yampa coal field, northwestern Colorado, in Field Guides: Geological Society of America, v. 1, p. 115–133.

Butler, D.L, Wright, W.G., Stewart, K.C., Osmundson, B.C., Krueger, R.P., and Crabtree, D.W., 1996, Detailed study of selenium and other constituents in water, bottom sediment, soil, alfalfa, and biota associated with irrigation drainage in the Uncompahgre Project area and in the Grand Valley, west-central Colorado, 1991–93: U.S. Geological Survey Water-Resources Investigations Report 96–4138, 136 p.

Campbell, D.H., and Turk, J.T., 1989, Effects of anions in snowmelt on acid neutralization in a watershed in the Front Range of Colorado: EOS Transactions, v. 70, p. 1,123.

Chigbu, Paulinus, and Sobeler, Dmitri, 2007, Bacteriological analysis of water, chap. 4 *of* Nollet, L.M.L., ed., Handbook of water analysis: Boca Raton, Fla., CRC Press, p. 97–134.

Childress, C.J.O., Foreman, W.T., Connor, B.F., and Maloney, T.J., 1999, New reporting procedures based on long-term method detection levels and some considerations for interpretations of water-quality data provided by the U.S. Geological Survey National Water Quality Laboratory: U.S. Geological Survey Open-File Report 99–193, 19 p., accessed September 2012, at *http://water.usgs.gov/owq/OFR_99-193/index.html.*

Colorado Department of Public Health and Environment, 2008, Seneca Coal Company proponent prehearing statement: Colorado Department of Public Health and Environment Water Quality Control Commission [variously paged].

Colorado Department of Public Health and Environment, 2009a, Water Quality Control Commission Regulation 31—The basic standards and methodologies for surface water: Colorado Department of Public Health and Environment Water Quality Control Commission, 186 p.

Colorado Department of Public Health and Environment, 2009b, Water Quality Control Commission Regulation 41—The basic standards for ground water: Colorado Department of Public Health and Environment Water Quality Control Commission, 64 p., accessed September 2012, at *http://www.colorado.gov/cs/Satellite?blobcol=urldata&blobheadername1=Content-Disposition&blobheadername2=Content-Type&blobheadervalue1=inline%3B+filename%3D%22Regulation+41.pdf%22&blobheadervalue2=application%2Fpdf&blobkey=id&blobtable=MungoBlobs&blobwhere=1251810001086&ssbinary=true.*

Colorado Department of Public Health and Environment, 2010, Water Quality Control Commission Regulations 33—Classification and numeric standards for Upper Colorado River Basin and North Platte River (planning region 12) and Regulation 33 tables: Colorado Department of Public Health and Environment Water Quality Control Commission, 64 p.

Colorado Department of Public Health and Environment, 2012, Colorado's section 303(d) list of impaired waters and monitoring and evaluation list [Regulation #93]: Colorado Department of Public Health and Environment Water Quality Control Commission, 97 p., accessed September 2012, at *http://www.colorado.gov/cs/Satellite?blobcol=urldata&blobheadername1=Content-Disposition&blobheadername2=Content-Type&blobheadervalue1=inline%3B+filename%3D%22Regulation+93.pdf%22&blobheadervalue2=application%2Fpdf&blobkey=id&blobtable=MungoBlobs&blobwhere=1251810003598&ssbinary=true.*

Colorado Geological Survey, 2005, Colorado coal—Energy security for the future [Rock Talk]: Denver, Colorado Geological Survey, v. 8, no. 2, 12 p, accessed September 2012, at *http://geosurvey.state.co.us/pubs/Documents/rtv8n21.pdf*.

Colorado Water Conservation Board, 2009, Yampa/White River basin information, Colorado's Decision Support System: Colorado Water Conservation Board [variously paged], accessed July 2010, at *ftp://dwrftp.state.co.us/cdss/swm/in/YampaBasinInfo_20091019.pdf*. (Also available at *http://cdss.state.co.us/basins/pages/YampaWhite.aspx*).

Colson, C.T., 1969, Stratigraphy and production of the Tertiary formations in the Sand Wash and Washakie basins, *in* Symposium on Tertiary rocks of Wyoming—Twenty-first Annual Field Conference Guidebook, 1969: Wyoming Geological Association, p. 121–128.

Covay, K.J., and Tobin, R.L., 1981, Quality of ground water in Routt County, Northwestern Colorado: U.S. Geological Survey Water-Resources Investigations Report Open-File Report 80–956, 38 p.

Driver, N.E., Norris, J.M., Kuhn, Gerhard, and others, 1984, Hydrology of Area 53, northern Great Plains and Rocky Mountain coal provinces, Colorado, Wyoming, and Utah: U.S. Geological Survey Open-File Report 83–765, 93 p.

Frazier, Deborah, 2000, Colorado's hot springs: Boulder, Colo., Pruett Publishing Company, p. 11–14.

GEI Consultants, Inc., 2007, Yampa River 2007 benthic invertebrate and water quality sampling: Littleton, Colo., GEI Consultants, Inc. [variously paged], accessed June 2012, at *http://steamboatsprings.net/sites/default/files/2007/12/01/Yampa_River__2007_Report.pdf*.

GEI Consultants, Inc., 2008, Yampa River 2008 habitat, benthic invertebrate, and water quality sampling: Littleton, Colo., GEI Consultants, Inc. [variously paged], accessed June 2012, at *http://media.steamboatpilot.com/news/documents/2009/02/11/Yampa_River_Report_2008.pdf*.

Helsel, D.R., 2005, Nondetects and data analysis—Statistics for censored environmental data: Hoboken, N.J., John Wiley, 250 p.

Helsel, D.R., and Hirsch, R. M., 2002, Statistical methods in water resources: U.S. Geological Survey, Techniques of Water-Resources Investigations, book 4, chap. A3, 522 p., accessed September 2010, at *http://water.usgs.gov/pubs/twri/twri4a3/*.

Hem, J.D., 1992, Study and interpretation of the chemical characteristics of natural waters: U.S. Geological Survey Water-Supply Paper 2254, 263 p., accessed June 2010, at *http://pubs.usgs.gov/wsp/wsp2254/*.

Hettinger, R.D., and Kirschbaum, M.A., 2002, Stratigraphy of the Upper Cretaceous Mancos Shale (upper part) and Mesaverde Group in the southern part of the Uinta and Piceance basins, Utah and Colorado: U.S. Geological Survey pamphlet for Geologic Investigations Series I–2764, 21 p, accessed June 2012, at *http://pubs.usgs.gov/imap/i-2764/i-2764_508.pdf*.

High Plains Regional Climate Center, 2010, Historical climate data summaries: Lincoln, Nebr., High Plains Regional Climate Center, accessed October 2009, at *http://www.hprcc.unl.edu/data/historical/*.

Hirsch, R.M., Slack, J.R., and Smith, R.A., 1982, Techniques of trend analysis for monthly water quality data: Water Resources Research, v. 18, p.107–121.

Johnson, E.A., Roberts, L.N.R., Brownfield, M.E., and Mercier, T.J., 2000, Geology and resource assessment of the middle and upper coal groups of the Yampa coal field, northwestern Colorado, *in* Kitschbaum, M.A., Roberts, L.N.R., and Biewick, L.R.H., eds., Geologic assessment of coal in the Colorado Plateau—Arizona, Colorado, New Mexico, and Utah: U.S. Geological Survey Professional Paper 1625–B, chap. P, p. P1–P65, accessed September 2010, at *http://pubs.usgs.gov/pp/p1625b/Reports/Chapters/Chapter_P.pdf*.

Jurgens, B.C., McMahon, P.B., Chapelle, F.H., and Eberts, S.M., 2009, An Excel® workbook for identifying redox processes in ground water: U.S. Geological Survey, Open-File Report 2009–1004, accessed December 2010, at *http://pubs.usgs.gov/of/2009/1004/*.

Kuhn, Gerhard, Stevens, M.R., and Elliott, J.G., 2003, Hydrology and water quality of Elkhead Creek and Elkhead Reservoir near Craig, Colorado, July 1995–September 2001: U.S. Geological Survey, Water-Resources Investigations Report 03–4220, 63 p., accessed September 2010, at *http://pubs.usgs.gov/wri/wri034220/pdf/508Kuhn.pdf*.

LaMotte, Andrew, 2008, National Land Cover Dataset 2001 (NLCD01) Tile 1, Northwest United States, NLCD01–1: U.S. Geological Survey [digital data], accessed February 2009, at *http://water.usgs.gov/GIS/metadata/usgswrd/XML/nlcd01_1.xml*.

Lund, J.W., 2006, Steamboat Springs, Colorado: Geo-Heat Center Quarterly Bulletin, v. 27, no. 3, p. 7–8, accessed September 2010, at *http://geoheat.oit.edu/bulletin/bull27-3/bull27-3-all.pdf*.

Lutgens, F.K., Tarbuck, E.J., Knapp, Jessica, Tasa, Dennis, and Richardson, Randall, 2012, Essentials of geology (11th ed.): Upper Saddle River, N.J., Prentice Hall.

Mast, M.A., 2007, Assessment of historical water-quality data for National Park units in the Rocky Mountain network through 2004: U.S. Geological Survey Scientific Investigations Report 2007–5147, 80 p., accessed August 2010, at *http://pubs.usgs.gov/sir/2007/5147/pdf/SIR2007-5147.pdf*.

Mast, M.A., Campbell, D.H., and Ingersoll, G.P, 2005, Effects of emission reductions at the Hayden Powerplant on precipitation, snowpack, and surface-water chemistry in the Mount Zirkel Wilderness Area, Colorado, 1995–2003: U.S. Geological Survey Scientific Investigations Report 2005–5167, 32 p., accessed October 2010, at *http://pubs.usgs.gov/sir/2005/5167/pdf/SIR2005-5167.pdf*.

Mast, M.A., Turk, J.T., Clow, D.W., and Campbell, D.H., 2011, Response of lake chemistry to changes in atmospheric deposition and climate in three high-altitude wilderness areas in Colorado: Biogeochemistry, v. 103, p. 27–43.

Mehls, S.F., and Mehls, C.D., 1991, Routt and Moffat Counties, Colorado, Coal mining historic context: Lafayette, Colo., Western Historical Associates, Inc., 93 p., accessed August 2010, at *http://coloradohistory-oahp.org/publications/pubs/620.pdf*. (Also available at *http://www.historycolorado.org/sites/default/files/files/OAHP/crforms_edumat/pdfs/620.pdf*).

Montgomery Watson Harza, 2002, Yampa basin watershed plan: Steamboat Springs, Colo., Montgomery Watson Harza [variously paged], accessed October 2009, at *http://www.cdphe.state.co.us/op/wqcc/Resources/208plans/208planfinal.pdf*.

Mueller, D.K., Hamilton, P.A., Helsel, D.R., Hitt, K.J., and Ruddy, B.C., 1995, Nutrients in groundwater and surface water of the United States—An analysis of data through 1992: U.S. Geological Survey Water-Resources Investigations Report 95–4031, 74 p.

Murphy, Sheila, 2007a, General information on phosphorus: Boulder, Colo., Boulder Area Sustainability Information Network, accessed September, 2012, at *http://bcn.boulder.co.us/basin/data/NEW/info/TP.html*.

Murphy, Sheila, 2007b, General information on solids: Boulder, Colo., Boulder Area Sustainability Information Network, accessed September, 2012, at *http://bcn.boulder.co.us/basin/data/NEW/info/TSS.html*.

National Park Service, 2001, Baseline water quality data inventory and analysis, Rocky Mountain National Park: National Park Service, 1,871 p., accessed July 2010, at *http://nrdata.nps.gov/romo/nrdata/water/baseline_wq/docs/ROMOWQAA.pdf*.

Robson, S.G., and Stewart, Michael, 1990, Geohydrologic evaluation of the upper part of the Mesaverde Group, northwestern Colorado: U.S. Geological Survey Water-Resources Investigations Report 90–4020, 120 p.

Rounds, S.A., 2006, Alkalinity and acid neutralizing capacity: U.S. Geological Survey Techniques of Water Resources Investigations, book 9 [National Field Manual], chap. A6.6, accessed April 2008, at *http://water.usgs.gov/owq/FieldManual/Chapter6/section6.6/*.

Runkel, R.L., Kimball, B.A., Walton-Day, Katherine, and Verplank, P.L., 2007, A simulation-based approach for estimating premining water quality—Red Mountain Creek, Colorado: Applied Geochemistry, v. 22, no. 8, p. 1,899–1,918.

Santore, R.C., Di Toro, D.M., Paquin, R.C., Allen, H.E., and Meyer, J.S., 2001, A biotic ligand model for the acute toxicity of metals, 2—Application to acute copper toxicity in freshwater fish and Daphnia: Environmental Toxicology and Chemistry, v. 20, no. 10, p. 2,397–2,402.

Schertz, T.L., Alexander, R.B., and Ohe, D.J., 1991, The computer program EStimate TREND (ESTREND), a system for the detection of trends in water-quality data: U.S. Geological Survey Water-Resources Investigations Report 91–4040, 72 p., accessed September 2010, at *http://water.usgs.gov/pubs/wri/wri91-4040/*.

State Demography Office, 2010a, Colorado jobs by sectors: Colorado Department of Local Affairs, Division of Local Government, accessed August 2010, at *https://dola.colorado.gov/demog_webapps/jsn_parameters.jsf*.

State Demography Office, 2010b, Population totals for Colorado counties: Colorado Department of Local Affairs, Division of Local Government, accessed August 2010, at *http://www.colorado.gov/cs/Satellite/DOLA-Main/CBON/1251590805419*.

Stephens, D.W., and Waddell, Bruce, 1998, Selenium sources and effects on biota in the Green River basin of Wyoming, Colorado, and Utah, in Frankenberger, W.T., Jr., and Engberg, R.A., eds., Environmental chemistry of selenium: New York, Marcel Dekker, p. 183–203.

Stoeser, D.B., Green, G.N., Morath, L.C., Heran, W.D., Wilson, A.B., Moore, D.W., and Van Gosen, B.S., 2005, Preliminary integrated geologic map databases for the United States, Central States—Montana, Wyoming, Colorado, New Mexico, North Dakota, South Dakota, Nebraska, Kansas, Oklahoma, Texas, Iowa, Missouri, Arkansas, and Louisiana: U.S. Geological Survey Open-File Report 2005–1351, version 1.2, updated December 2007, accessed June 2012, at *http://pubs.usgs.gov/of/2005/1351/*.

Topper, Ralf, Spray, K.L, Bellis, W.H., Hamilton, J.L., and Barkmann, P.E., 2003, Ground water atlas of Colorado: Colorado Geological Survey [variously paged], accessed October 2010, at *http://geosurveystore.state.co.us/p-862-ground-water-atlas-of-colorado.aspx*.

Turk, J.T., and Spahr, N.E., 1991 Rocky Mountains, *in* Charles, D.F., ed., Acidic deposition and aquatic ecosystems: New York, Springer, p. 471–499.

Tweto, Ogden, 1979, Geologic map of Colorado: U.S. Geological Survey Special Map, scale 1:500,000.

U.S. Environmental Protection Agency, 2000, Ambient water quality criteria recommendations—Information supporting the development of state and tribal nutrient criteria [for] rivers and streams in Ecoregion II: U.S. Environmental Protection Agency Report EPA 822–B–00–015 [variously paged], accessed November 2009, at *http://www.epa.gov/waterscience/criteria/nutrient/ecoregions/rivers/rivers_2.pdf*.

U.S. Environmental Protection Agency, 2010a, Map of radon zones for Colorado: U.S. Environmental Protection Agency, accessed May 2010, at *http://www.epa.gov/radon/states/colorado.html#zone%20map*.

U.S. Environmental Protection Agency, 2010b, Safe Drinking Water Information System, List of water systems in SDWIS [Routt County, Colorado]: U.S. Environmental Protection Agency, accessed May 2010, at *http://oaspub.epa.gov/enviro/sdw_query_v2.get_list?wsys_name=&fac_search=fac_beginning&fac_county=ROUTT&pop_serv=500&pop_serv=3300&pop_serv=10000&pop_serv=100000&pop_serv=100001&sys_status=active&pop_serv=&wsys_id=&fac_state=CO&last_fac_name=&page=1&query_results=&total_rows_found=*.

U.S. Environmental Protection Agency, 2012a, Radionuclides in drinking water: U.S. Environmental Protection Agency, accessed September 2012, at *http://water.epa.gov/lawsregs/rulesregs/sdwa/radionuclides/index.cfm*.

U.S. Environmental Protection Agency, 2012b, Radon: U.S. Environmental Protection Agency, accessed September 2012, at *http://www.epa.gov/rpdweb00/radionuclides/radon.html*.

U.S. Geological Survey, 2010, Streamstats—A water resources web application, accessed October 2009, at *http://streamstatsags.cr.usgs.gov/co_ss/*.

U.S. Geological Survey, 1991, Contamination of dissolved trace-element data—Present understanding, ramifications, and issues that require resolution: Office of Water Quality Technical Memorandum 91.10, U.S. Geological Survey, accessed June 2012, at *http://water.usgs.gov/admin/memo/QW/qw91.10.html*.

Water Quality Control Division, 2002, Policy for characterizing ambient water quality for use in determining water quality standards based effluent limits: [Colorado] Water Quality Control Division, 2 p., accessed October 2012, at *http://www.colorado.gov/cs/Satellite?blobcol=urldata&blobheadername1=Content-Disposition&blobheadername2=Content-Type&blobheadervalue1=inline%3B+filename%3D%22WQP-19.pdf%22&blobheadervalue2=application%2Fpdf&blobkey=id&blobtable=MungoBlobs&blobwhere=1251816809833&ssbinary=true*.

Wentz, D.A., 1974, Effect of mine drainage in the quality of streams in Colorado, 1971–72: Denver, Colorado Water Conservation Board, 117 p., accessed July 2010, at *http://co.water.usgs.gov/publications/non-usgs/CWR_circ21.pdf*.

Wetzel, R.G., 1983, Limnology: Fort Worth, Saunders College Publishing [variously paged].

Appendix 1.   U.S. Environmental Protection Agency STORET edit-checking procedure of low and high values for selected water-quality properties and constituents from the Upper Yampa River watershed water-quality database.

[C, degrees Celsius; µS/cm, microsiemens per centimeter at 25 degrees Celsius; mg/L, milligrams per liter; $CaCO_3$, calcium carbonate; HCO_3^-, bicarbonate; N, nitrogen; PO_4, phosphate; P, phosphorus; µg/L, micrograms per liter; m-FC, membrane-Fecal Coliform; col/100 mL, colonies per 100 milliliters; NO_3, nitrate. Data are from National Park Service (2001)]

Parameter code[1]	Parameter code name for water-quality property or constituent (reporting units)[1]	Low value	High value
P00010	Water temperature (°C)	−2	37.0
P00094	Specific conductance (µS/cm)	1.0	60,000.0
P00095	Specific conductance (µS/cm)	1.0	60,000.0
P00300	Oxygen, dissolved (mg/L)	0.0	30.0
P00400	pH (standard units)	0.9	12.0
P00403	pH, lab (standard units)	0.9	12.0
P00410	Alkalinity, total (mg/L as $CaCO_3$)	0.0	1,000.0
P00440	Bicarbonate (mg/L as HCO_3^-)	0.0	450.0
P00600	Nitrogen, total (mg/L as N)	0.0	100.0
P00610	Ammonia, total (mg/L as N)	0.0	20.0
P00615	Nitrite, total (mg/L as N)	0.0	5.0
P00625	Ammonia plus organic nitrogen, total (mg/L as N)	0.0	50.0
P00630	Nitrite plus nitrate, total (mg/L as N)	0.0	55.0
P00650	Phosphate, total (mg/L as PO_4)	0.0	30.0
P00665	Phosphorus, total (mg/L as P)	0.0	10.0
P00900	Hardness, total (mg/L as $CaCO_3$)	0.0	5,000.0
P00915	Calcium, dissolved (mg/L)	0.0	1,000.0
P00925	Magnesium, dissolved (mg/L)	0.0	1,000.0
P00930	Sodium, dissolved (mg/L)	0.0	5,000.0
P00935	Potassium, dissolved (mg/L)	0.0	1,000.0
P00940	Chloride, dissolved (mg/L)	0.0	22,000.0
P00945	Sulfate, total (mg/L)	0.0	2,500.0
P00946	Sulfate, dissolved (mg/L)	0.0	2,500.0
P00950	Fluoride, dissolved (mg/L)	0.0	15.0
P00955	Silica, dissolved (mg/L)	0.0	2,000.0
P01000	Arsenic, dissolved (µg/L)	0.0	5,000.0
P01002	Arsenic, total (µg/L)	0.0	5,000.0
P01005	Barium, dissolved (µg/L)	0.0	2,000.0
P01007	Barium, total (µg/L)	0.0	2,000.0
P01010	Beryllium, dissolved (µg/L)	0.0	2,000.0
P01012	Beryllium, total (µg/L)	0.0	2,000.0
P01020	Boron, dissolved (µg/L)	0.0	5,000.0
P01022	Boron, total (µg/L)	0.0	5,000.0
P01025	Cadmium, dissolved (µg/L)	0.0	500.0
P01027	Cadmium, total (µg/L)	0.0	500.0
P01030	Chromium, dissolved (µg/L)	0.0	2,000.0
P01032	Chromium, hexavalent (µg/L)	0.0	2,000.0
P01034	Chromium, total (µg/L)	0.0	2,000.0
P01040	Copper, dissolved (µg/L)	0.0	2,000.0
P01042	Copper, total (µg/L)	0.0	5,000.0
P01045	Iron, total (µg/L)	0.0	56,000.0
P01046	Iron, dissolved (µg/L)	0.0	56,000.0
P01049	Lead, dissolved (µg/L)	0.0	1,000.0
P01051	Lead, total (µg/L)	0.0	1,000.0

Appendix 1. U.S. Environmental Protection Agency STORET edit-checking procedure of low and high values for selected water-quality properties and constituents from the Upper Yampa River watershed water-quality database.—Continued

[C, degrees Celsius; μS/cm, microsiemens per centimeter at 25 degrees Celsius; mg/L, milligrams per liter; $CaCO_3$, calcium carbonate; HCO_3^-, bicarbonate; N, nitrogen; PO_4, phosphate; P, phosphorus; μg/L, micrograms per liter; m-FC, membrane-Fecal Coliform; col/100 mL, colonies per 100 milliliters; NO_3, nitrate. Data are from National Park Service (2001)]

Parameter code[1]	Parameter code name for water-quality property or constituent (reporting units)[1]	Low value	High value
P01055	Manganese, total (μg/L)	0.0	5,000.0
P01056	Manganese, dissolved (μg/L)	0.0	5,000.0
P01065	Nickel, dissolved (μg/L)	0.0	2,000.0
P01067	Nickel, total (μg/L)	0.0	2,000.0
P01075	Silver, dissolved (μg/L)	0.0	5,000.0
P01077	Silver, total (μg/L)	0.0	5,000.0
P01090	Zinc, dissolved (μg/L)	0.0	25,000.0
P01092	Zinc, total (μg/L)	0.0	25,000.0
P01105	Aluminum, total (μg/L)	0.0	20,000.0
P01106	Aluminum, dissolved (μg/L)	0.0	20,000.0
P01145	Selenium, dissolved (μg/L)	0.0	100.0
P22703	Uranium, natural, dissolved (μg/L)	0.0	500.0
P31613	Fecal coliform, membrane filter, m-FC agar, 44.5 °C, 24 hour (col/100 mL)	0.0	10,000,000.0
P31616	Fecal coliform, membrane filter, m-FC broth (col/100 mL)	0.0	10,000,000.0
P70300	Residue, total filtrable, dried at 180 °C (mg/L)	0.0	4,000.0
P70507	Phosphorus, orthophosphate, total (mg/L as P)	0.0	10.0
P71850	Nitrate nitrogen, total (mg/L as NO_3)	0.0	65.0
P71890	Mercury, dissolved (μg/L)	0.0	10.0
P71900	Mercury, total (μg/L)	0.0	10.0

[1]Parameter codes and parameter code names for water-quality properties or constituents are from the U.S. Geological Survey National Water Information System (NWIS) and the U.S. Environmental Protection Agency Data STOrage and RETrieval System (STORET) Data Warehouse.

Appendix 2. Selected Colorado Department of Public Health and Environment in-stream water-quality standards for stream segments in the Upper Yampa River watershed, Colorado

[mg/L, milligrams per liter; C, degrees Celsius; DM, daily maximum temperature; MWAT, maximum weekly average temperature; SMCL, secondary maximum contaminant level; N, nitrogen; MCL; maximum contaminant level; TVS, table value standard; P, phosphorus; µg/L, micrograms per liter; tr, trout; FRV, Final Residue Value; sc, sculpin; CaCO$_3$, calcium carbonate; $\geq$ greater than or equal to; col/100 mL, colonies per 100 milliliters. Water-quality standards are from Colorado Department of Public Health and Environment (2009a, 2010), unless otherwise stated. Descriptions of segments are in Colorado Department of Public Health and Environment (2010). Recommended total phosphorus levels are from U.S. Environmental Protection Agency (2000). Stream segments with no water-quality data in the Upper Yampa River watershed database or outside of the Upper Yampa River watershed boundary as defined for this report are not included in this table]

Physical property or constituent (units)	In-stream water-quality standard		Stream segment
	Type	Value[1]	
Physical properties			
pH (standard units)	Aquatic life	6.5–9.0	all
Dissolved oxygen (mg/L)	Aquatic life	6.0	1b, 2a–2c, 3–8, 11, 12, 13a–13c, 13f, 14, 20a
	Aquatic life	5.0	13d, 13e, 15
Water temperature (°C) (streams)	Aquatic life cold (June–Sept)	21.2 (acute, DM[2])	2a, 3, 5, 6, 8, 11, 13a, 20a
		17.0 (chronic, MWAT[3])	
	Aquatic life cold (Oct–May)	13.0 (acute, DM[2])	
		9.0 (chronic, MWAT[3])	
	Aquatic life cold (Apr–Oct)	23.8 (acute, DM[2])	2c, 4, 7, 12, 13b, 13c, 13f, 14
		18.2 (chronic, MWAT[3])	
	Aquatic life cold (Nov–Mar)	13.0 (acute, DM[2])	
		9.0 (chronic, MWAT[3])	
	Aquatic life warm (Mar–Nov)	28.6 (acute, DM[2])	13d, 13e, 15
		27.5 (chronic, MWAT[3])	
	Aquatic life warm (Dec–Feb)	14.3 (acute, DM[2])	
		13.7 (chronic, MWAT[3])	
Water temperature (°C) (lakes and reservoirs)	Aquatic life cold (Apr–Dec)	21.2 (acute, DM[2])	1b, 2b
		17.0 (chronic, MWAT[3])	
	Aquatic life cold (Jan–Mar)	13.0 (acute, DM[2])	
		9.0 (chronic, MWAT[3])	
	Aquatic life cold (Apr–Dec)	23.8 (acute, DM[2])	2b[4,5]
		18.2 (chronic, MWAT[3])	
	Aquatic life cold (Jan–Mar)	13.0 (acute, DM[2])	
		9.0 (chronic, MWAT[3])	
Dissolved solids and major ions			
Dissolved solids (mg/L)	Water supply (SMCL)[6]	500	1b, 2a–2c, 3, 4, 6, 8, 13a, 13c (June–Feb), 13f, 14, 15, 20a
Sulfate, unfiltered (mg/L)	Water supply (SMCL)	250	1b, 2a–2c, 3, 4, 6, 8, 13a, 13c (June–Feb), 13f, 14, 15, 20a
Chloride, unfiltered (mg/L)	Water supply (SMCL)	250	1b, 2a–2c, 3, 4, 6, 8, 13a, 13c (June–Feb), 13f, 14, 15, 20a
Nutrients			
Nitrite, unfiltered (mg/L as N)	Aquatic life	0.05	all
Nitrate, unfiltered (mg/L as N)	Water supply (MCL)	10	1b, 2a–2c, 3, 4, 6, 8, 13a, 13c (June–Feb), 13f, 14, 15, 20a
	Agriculture	100	5, 7, 11, 12, 13b, 13c (March–May), 13d, 13e
Total ammonia (mg/L as N)	Aquatic life	TVS (acute, chronic, or none)	1b, 2a–2c, 3, 5–8, 13a–13f, 14, 15, 20a
Total phosphorus (mg/L as P)	Recommended[7]	0.05	2a, 3, 15 (Streams that flow directly into lake or reservoir)
		0.1	2a, 2c, 3, 4, 6, 8, 13a, 13c (June–Feb), 13f, 14, 15, 20a (Streams that do not flow directly into lake or reservoir)

Appendix 2. Selected Colorado Department of Public Health and Environment in-stream water-quality standards for stream segments in the Upper Yampa River watershed, Colorado.—Continued

[mg/L, milligrams per liter; °C, degrees Celsius; DM, daily maximum temperature; MWAT, maximum weekly average temperature; SMCL, secondary maximum contaminant level; N, nitrogen; MCL; maximum contaminant level; TVS, table value standard; P, phosphorus; µg/L, micrograms per liter; tr, trout; FRV, Final Residue Value; sc, sculpin; $CaCO_3$, calcium carbonate; $\geq$, greater than or equal to; col/100 mL, colonies per 100 milliliters. Water-quality standards are from Colorado Department of Public Health and Environment (2009a, 2010), unless otherwise stated. Descriptions of segments are in Colorado Department of Public Health and Environment (2010). Recommended total phosphorus levels are from U.S. Environmental Protection Agency (2000). Stream segments with no water-quality data in the Upper Yampa River watershed database or outside of the Upper Yampa River watershed boundary as defined for this report are not included in this table]

Physical property or constituent (units)	In-stream water-quality standard Type	Value[1]	Stream segment
		Trace elements	
Arsenic, dissolved (µg/L)	Aquatic life	340 (acute)	all
Arsenic, total recoverable (µg/L)	Aquatic life	0.02 (chronic)	1b, 2a–2c, 3, 6, 8, 13a, 13c (June–Feb), 13f, 14, 15, 20a
		0.02–10 (chronic)	[8]4
Boron, dissolved (µg/L)	Agriculture	7.6 (chronic)	5, 7, 13b, 13c (March–May)
		100 (chronic)	[8]11, 12, 13d, 13e
Cadmium, dissolved (µg/L)	Aquatic life	750	all
		TVS(tr) (acute)	1b, 2a–2c, 3, 5–8, 13a–3c, 13f, 14, 15, 20a
		TVS (acute)	13d, 13e
		TVS (chronic)	1b, 2a–2c, 3, 5–8, 13a–13f, 14, 15, 20a
Cadmium, total recoverable (µg/L)	Aquatic life	5 (acute)	[8]4
		10 (chronic)	[8]11, 12
Chromium, dissolved (µg/L)	Aquatic life	TVS (16 acute, 11 chronic)	1b, 2a–2c, 3, 5–8, 13a–13e, 14, 15, 20a
Chromium, total recoverable (µg/L)	Aquatic life	50 (acute)	[8]4
		100 (chronic)	11, 12
Copper, dissolved (µg/L)	Aquatic life	TVS (acute, chronic)	1b, 2a–2c, 3, 5–8, 13a–13e, 14, 15, 20a
Copper, total recoverable (µg/L)	Aquatic life	200 (chronic)	[8]4
		200 (acute)	11, 12
Iron, dissolved (µg/L)	Water supply (SMCL)	300 (chronic)	1b, 2a–2c, 3, 4, 6, 8, 13a, 13c (June–Feb), 13f, 14, 15, 20a
Iron, total recoverable (µg/L)	Aquatic life	1,000 (chronic)	1b, 2a–2c, 3, 5–8, 13a, 13b (not Middle Creek), 13c, 13f, 14, 15, 20a
		1,035 (chronic)	13b (Middle Creek)
		existing quality (chronic)	13d, 13e
Lead, dissolved (µg/L)	Aquatic life	TVS (acute, chronic)	1b, 2a–2c, 3, 5–8, 13a–13e, 14, 15, 20a
Lead, total recoverable (µg/L)	Aquatic life	50 (acute)	[8]4
		100 (chronic)	11, 12
Manganese, dissolved (µg/L)	Water supply (SMCL)	TVS (acute, chronic)	1b, 2a–2c, 3, 5–8, 13a–13f, 14, 15, 20a
		50 (chronic)	1b, 2a–2c, 3, 6, 8, 13a, 13c (June–Feb), 13f, 14, 15, 20a
Manganese, total recoverable (µg/L)	Aquatic life	TVS (acute, chronic)	[8]4
		200 (chronic)	11, 12
Mercury, total, dissolved (µg/L)	Water supply (SMCL) FRV[9]	50 (chronic)	[8]4
		0.01 (chronic)	1b, 2a–2c, 3, 5–8, 13a–13e, 14, 15, 20a
Mercury, total recoverable (µg/L)	Aquatic life	2.0 (acute)	4
Nickel, dissolved (µg/L)	Aquatic life	TVS (acute, chronic)	1b, 2a–2c, 3, 5–8, 13a–13e, 14, 15, 20a
		100 (chronic)	[8]4
Nickel, total recoverable (µg/L)	Aquatic life	200 (chronic)	11, 12
Selenium, dissolved (µg/L)	Aquatic life	TVS (18.4 acute, 4.6 chronic)	1b, 2a–2c, 3, 5–8, 13a–13e, 14, 15, 20a

Appendix 2. Selected Colorado Department of Public Health and Environment in-stream water-quality standards for stream segments in the Upper Yampa River watershed, Colorado.—Continued

[mg/L, milligrams per liter; C, degrees Celsius; DM, daily maximum temperature; MWAT, maximum weekly average temperature; SMCL, secondary maximum contaminant level; N, nitrogen; MCL; maximum contaminant level; TVS, table value standard; P, phosphorus; μg/L, micrograms per liter; tr, trout; FRV, Final Residue Value; sc, sculpin; $CaCO_3$, calcium carbonate; ≥, greater than or equal to; col/100 mL, colonies per 100 milliliters. Water-quality standards are from Colorado Department of Public Health and Environment (2009a, 2010), unless otherwise stated. Descriptions of segments are in Colorado Department of Public Health and Environment (2010). Recommended total phosphorus levels are from U.S. Environmental Protection Agency (2000). Stream segments with no water-quality data in the Upper Yampa River watershed database or outside of the Upper Yampa River watershed boundary as defined for this report are not included in this table]

Physical property or constituent (units)	In-stream water-quality standard		Stream segment
	Type	Value[1]	
Trace elements—Continued			
Silver, dissolved (μg/L)	Aquatic life	TVS (acute)	1b, 2a–2c, 3, 5–8, 13a–13f, 14, 15, 20a
		TVS(tr) (chronic)	1b, 2a–2c, 3, 5–8, 13a–13c, 13f, 14, 15, 20a
		TVS (chronic)	13d, 13e
			[8]4
Silver, total recoverable (μg/L)	Aquatic life	100 (acute)	1b, 2a–2c, 3, 5–8, 13a–13f, 14, 15, 20a
Zinc, dissolved (μg/L)	Aquatic life	TVS (acute)	2a, 2c, 3, 8, 13a (hardness less than 113 mg/L $CaCO_3$)
		TVS(sc) (chronic)	1b, 2b, 5–7, 13b–13f, 14, 15, 20a; 2a, 2c, 3, 8, 13a (hardness ≥ 113 mg/L $CaCO_3$)
		TVS (chronic)	[8]4
Zinc, total recoverable (μg/L)	Aquatic life	2,000 (acute)	[10]4, 11, 12
		2,000 (chronic)	
Radiochemical			
Uranium, natural, dissolved (μg/L)	Water supply	[11]30	all
Coliform bacteria			
Escherichia coli (col/100 mL)	Recreation	[12]126/100	1b, 2a–2c, 3, 6, 8, 13a–13d, 13f, 14, 15, 20a
		[12]205/100	[8]5, 7
		[12]630/100	[8]4, 11, 12, 13e

[1]Table value standards for trace elements vary with hardness (Colorado Department of Public Health and Environment, 2009a, 2010).

[2]Daily maximum temperature (DM) is the highest 2-hour average water temperature recorded during a given 24-hour period (Colorado Department of Public Health and Environment, 2009a).

[3]Maximum weekly average temperature (MWAT) is the maximum average of multiple, equally spaced, daily temperatures over 7 consecutive days with a minimum of three data points spaced equally throughout the day (Colorado Department of Public Health and Environment, 2009a).

[4]Temperature standard applies to lakes and reservoirs with a surface area greater than or equal to 100 acres surface area (Colorado Department of Public Health and Environment, 2009a).

[5]April–December temperature WAT (weekly average temperature) for stream segment 2b: Stagecoach Reservoir, 21.4 C; Steamboat Lake, 21.6 C (Colorado Department of Public Health and Environment, 2010).

[6]For stream segments with dissolved solids data, the standard has been applied to stream segments that have standards for unfiltered sulfate and chloride.

[7]For stream segments with total phosphorus data, the recommendation has been applied to stream segments that have a standard for unfiltered nitrate.

[8]No data were available for the constituent of interest in the steam segment(s) listed.

[9]Final residue value (FRV) is the maximum allowed concentration of total mercury in water that will present bioaccumulation or bioconcentration of methylmercury in edible fish tissue (Colorado Department of Public Health and Environment, 2009a).

[10]No data for total recoverable zinc were available for segment 4.

[11]Uranium level in surface water used for water supply shall be maintained at the lowest practical level. The maximum allowed concentration in water used for supply is 30 micrograms per liter, unless naturally occurring concentrations are greater (Colorado Department of Public Health and Environment, 2009a).

[12]Attainment of the Colorado Department of Public Health and Environment water-quality standard is based on the geometric mean of the data.

Appendix 3.   Description of selected stream sampling sites in the Upper Yampa River watershed, Colorado, with type of water-quality data collected, period of water-quality record, and number of samples collected, 1975 through 2009

[No , number; USFS, U.S. Department of the Agriculture Forest Service; PP, physical properties; D, discharge; CB, coliform bacteria; SS, suspended sediment; CDPHE, Colorado Department of Public Health and Environment; DS, dissolved solids; MI, major ions; N, nutrients, TE, trace elements; CODOW, Colorado Division of Wildlife Riverwatch Program; USGS, U.S. Geological Survey; DOC, dissolved organic carbon; CSS, City of Steamboat Springs; U, uranium; USEPA, U.S. Environmental Protection Agency. Subwatershed definitions: Yampa River subwatershed 1, Yampa River and tributaries upstream from Chuck Lewis State Wildlife Area; Yampa River subwatershed 2, Yampa River and tributaries from Chuck Lewis State Wildlife Area to Elk River confluence; Elk River subwatershed, Elk River and tributaries; Yampa River subwatershed 3, Yampa River and tributaries from Elk River confluence to Town of Hayden; Yampa River subwatershed 4, Yampa River and tributaries from Town of Hayden to Elkhead Creek confluence; Elkhead Creek subwatershed, Elkhead Creek and tributaries. Sites with the same site number are considered to be at the same location. Sites with name in bold are streamgage stations; see table 1 for additional information on these sites. The Upper Yampa River water-quality database is available at *http //rmgsc.cr.usgs.gov/cwqdr/Yampa/index.shtml*]

Site no. (see figure 3)	Site name in Upper Yampa River watershed water-quality database	Source of data	Site identifier	Latitude
1	BEAR R. #2 13 MI SW YAMPA	USFS	11057702	40.05
2	BEAR R. #3 10MI SW YAMPA	USFS	11057703	40.067
3	BEAR R. @ RD. 8	CDPHE	12898	40.157
4	BEAR RIVER #1	USFS	11057701	40.033
5	Bear River Miller	CODOW	CDOWRW-6	40.08
6	BEAR RIVER NEAR TOPONAS, CO	USGS	09236000	40.044
7	BEAVER CREEK NEAR HAHNS PEAK, CO.	USGS	404610106545600	40.769
8	BTKN-DWN [Butcherknife Creek 15m above Yampa St]	CSS	BTKN-DWN	40.484
9	BURGESS C TRIB BL SKI AREA NR STEAMBOAT SPGS, CO	USGS	402745106473600	40.462
10	BURGESS CK NEAR MOUTH @ HWY 40	CDPHE	12893	40.452
11	BURGESS CREEK AB SKI AREA NR STEAMBOAT SPGS, CO.	USGS	402802106471000	40.467
12	BURGESS CREEK BL SKI AREA NR STEAMBOAT SPGS, CO.	USGS	402720106481500	40.456
13	BUTCHERKNIFE CREEK NR MOUTH AT STEAMBOAT SPG,CO.	USGS	402944106495900	40.496
14	CHENEY CREEK NEAR MILNER,CO.	USGS	402908107014000	40.486
15	CHIMNEY CK @ RD. 8	CDPHE	12899	40.158
16	CHIMNEY CREEK AT TRAPPER, CO.	USGS	400612106524800	40.103
17	COW CR. NR. STEAMBOAT SPRINGS,CO.	USGS	402836106550100	40.477
18	CREEK ROUTT	USGS	403333106504900	40.559
19	DEEP CREEK AT HAHNS PEAK, CO.	USGS	404845106571400	40.812
20	DILL GULCH NEAR HAYDEN, CO	USGS	402605107181500	40.435
21	DILL GULCH TRIBUTARY 0.2MI AB MOUTH -S132	USGS	402558107182101	40.433
22	DRY CK @ HAYDEN	CDPHE	12852	40.492
22	DRY CREEK AT HWY 40 AT HAYDEN, CO	USGS	402938107160101	40.494
23	DRY CK AT HAYDEN NR YAMPA VALLEY AIRPORT	CDPHE	12852A	40.481
24	DRY CREEK ABOVE SEWAGE PLANT AT HAYDEN, CO.	USGS	402939107160100	40.494
25	DRY CREEK BELOW SEWAGE PLANT AT HAYDEN, CO.	USGS	402952107161600	40.498
26	ELK R.@ RD. 64A	CDPHE	12868	40.752
27	ELK R. NEAR MOUTH @ CR44	CDPHE	12860	40.546
28	ELK RIVER AB. GLEN EDEN	CDPHE	12865	40.716
29	ELK RIVER ABOVE CLARK, CO	USGS	09240900	40.743
30	ELK RIVER AT CLARK, CO.	USGS	09241000	40.717
31	ELK RIVER BELOW SOUTH FORK AT HINMAN PARK, CO.	USGS	404506106492800	40.752
32	ELK RIVER NEAR MILNER	CDPHE	000154	40.483
33	**ELK RIVER NEAR MILNER, CO.**	USGS	09242500	40.515
33	ELK R. @ RD. 42	CDPHE	12861	40.515
34	ELK RIVER NEAR MOUTH	USGS	402914106580400	40.487
34	ELK RIVER NEAR MOUTH AT US 40 BRIDGE, CO.	USGS	402913106580400	40.487
35	ELKHEAD CR 1.5 MILES BELOW NORTH FORK	CDPHE	ELKHEAD03	40.65
36	ELKHEAD CR AT HWY 40 BRIDGE	CDPHE	ELKHEAD01	40.667
37	ELKHEAD CR BELOW ELKHEAD RES	CDPHE	ELKHEAD02	40.55
38	ELKHEAD CREEK	USEPA	WCOP99-0512	40.660
39	ELKHEAD CREEK	USEPA	WCOP99-0565	40.620
40	**ELKHEAD CREEK ABOVE LONG GULCH, NEAR HAYDEN, CO**	USGS	09246200	40.592
40	ELKHEAD CREEK ABOVE ELKHEAD RESERVOIR, CO.	USGS	403530107191300	40.592
41	ELKHEAD CREEK BELOW ELKHEAD RESERVOIR, CO	USGS	403318107230100	40.555
42	ELKHEAD CREEK BELOW MAYNARD GULCH, NEAR CRAIG, CO	USGS	09246400	40.545
43	**ELKHEAD CREEK NEAR CRAIG, CO**	USGS	09246500	40.531
43	ELKHEAD CK NR CRAIG @ HWY 40	CDPHE	12840	40.531
43	ELKHEAD CREEK NEAR MOUTH	USGS	403152107260700	40.531

Appendix 3.   Description of selected stream sampling sites in the Upper Yampa River watershed, Colorado, with type of water-quality data collected, period of water-quality record, and number of samples collected, 1975 through 2009.—Continued

[No , number; USFS, U.S. Department of the Agriculture Forest Service; PP, physical properties; D, discharge; CB, coliform bacteria; SS, suspended sediment; CDPHE, Colorado Department of Public Health and Environment; DS, dissolved solids; MI, major ions; N, nutrients, TE, trace elements; CODOW, Colorado Division of Wildlife Riverwatch Program; USGS, U.S. Geological Survey; DOC, dissolved organic carbon; CSS, City of Steamboat Springs; U, uranium; USEPA, U.S. Environmental Protection Agency. Subwatershed definitions: Yampa River subwatershed 1, Yampa River and tributaries upstream from Chuck Lewis State Wildlife Area; Yampa River subwatershed 2, Yampa River and tributaries from Chuck Lewis State Wildlife Area to Elk River confluence; Elk River subwatershed, Elk River and tributaries; Yampa River subwatershed 3, Yampa River and tributaries from Elk River confluence to Town of Hayden; Yampa River subwatershed 4, Yampa River and tributaries from Town of Hayden to Elkhead Creek confluence; Elkhead Creek subwatershed, Elkhead Creek and tributaries. Sites with the same site number are considered to be at the same location. Sites with name in bold are streamgage stations; see table 1 for additional information on these sites. The Upper Yampa River water-quality database is available at *http //rmgsc.cr.usgs.gov/cwqdr/Yampa/index.shtml*]

Site no. (see figure 3)	Longitude	Subwatershed	Type of water-quality data collected[1]	Period of water-quality record (calendar year)	No. of samples
1	−107.067	Yampa River subwatershed 1	PP,D,CB,SS	1975	6
2	−107.033	Yampa River subwatershed 1	PP,D,CB,SS	1975, 1978–79	24
3	−106.902	Yampa River subwatershed 1	PP,D,DS,MI,N,TE	[2]2000–07	9
4	−107.117	Yampa River subwatershed 1	PP,CB,SS	1975	5
5	−106.97	Yampa River subwatershed 1	PP,D,TE	1990–99	131
6	−107.072	Yampa River subwatershed 1	PP,D,N,TE,DOC	[2]1975–86, 2005	46
7	−106.916	Elk River subwatershed	PP,D,N,TE,DOC	1975	1
8	−106.834	Yampa River subwatershed 2	PP,MI,N,TE,CB	2007–08	4
9	−106.794	Yampa River subwatershed 2	PP,D,N	1976	3
10	−106.810	Yampa River subwatershed 2	PP,D,DS,MI,N,TE,CB	1999	1
11	−106.787	Yampa River subwatershed 2	PP,D,N,CB,SS	1975–76	8
12	−106.805	Yampa River subwatershed 2	PP,D,CB,SS	1975–76	5
13	−106.821	Yampa River subwatershed 2	PP,D,N,CB	1975–76	5
14	−107.028	Yampa River subwatershed 3	D,CB	1975	1
15	−106.9	Yampa River subwatershed 1	PP,DS,MI,N,TE	[2]2001–07	17
16	−106.881	Yampa River subwatershed 1	PP,D,DS,MI,N,TE,DOC,CB,SS	1975–76	6
17	−106.918	Yampa River subwatershed 2	PP,D,DS,MI,N,TE,SS	1981–82, 2005	10
18	−106.848	Yampa River subwatershed 2	PP,MI,TE	1976	1
19	−106.954	Elk River subwatershed	PP,D,N,TE,DOC	1975	1
20	−107.305	Yampa River subwatershed 4	PP,D,DS,MI,N,TE,SS	1981–82	4
21	−107.306	Yampa River subwatershed 4	PP,D,DS,MI,N,TE	1982	1
22	−107.265	Yampa River subwatershed 4	PP,DS,MI,N,TE,U,CB	[2]2001–05	7
22	−107.267	Yampa River subwatershed 4	PP,D	2005	2
23	−107.236	Yampa River subwatershed 4	PP,MI,N,TE	2007	2
24	−107.268	Yampa River subwatershed 4	PP	1975	2
25	−107.272	Yampa River subwatershed 4	PP,D,N,CB	1975	4
26	−106.759	Elk River subwatershed	PP,DS,MI,N,TE,CB	2001–02	13
27	−106.909	Elk River subwatershed	PP,DS,MI,N,TE	1999	1
28	−106.916	Elk River subwatershed	PP,DS,MI,N,TE	1996–97	11
29	−106.855	Elk River subwatershed	PP,D	[2]1987–2003	99
30	−106.916	Elk River subwatershed	PP,D,DS,MI,N,TE,DOC,SS	[2]1975–2003	148
31	−106.825	Elk River subwatershed	PP,D,DS,MI,N,TE,DOC,CB,SS	1975–76	6
32	−106.979	Elk River subwatershed	PP,DS,MI,N,TE,U	1979–82	21
33	−106.954	Elk River subwatershed	PP,D,DS,MI,N,TE,DOC,CB,SS	1975–76, 1989–2005	180
33	−106.954	Elk River subwatershed	PP,DS,MI,N,TE,U,CB	[2]2001–07	25
34	−106.968	Elk River subwatershed	PP,D,DS,MI,N,DOC	1999–2000	2
34	−106.968	Elk River subwatershed	PP,N,CB	1975	2
35	−107.317	Elkhead Creek subwatershed	PP,DS	1979	1
36	−107.283	Elkhead Creek subwatershed	PP,DS,TE	1979	1
37	−107.383	Elkhead Creek subwatershed	PP,TE	1979	1
38	−107.291	Elkhead Creek subwatershed	PP,MI,N,TE,DOC	2000	2
39	−107.271	Elkhead Creek subwatershed	PP,MI,N,TE,DOC	2001	2
40	−107.321	Elkhead Creek subwatershed	PP,D,DS,MI,N,TE,DOC,CB,SS	1995–2004	143
40	−107.321	Elkhead Creek subwatershed	PP,D,DS,MI,N,TE,DOC,CB,SS	[2]1975–83	10
41	−107.384	Elkhead Creek subwatershed	PP,D,SS	[2]1997–2003	10
42	−107.398	Elkhead Creek subwatershed	PP,D,DS,MI,N,TE,DOC,CB,SS	1995–2005	143
43	−107.436	Elkhead Creek subwatershed	PP,D,DS,MI,N,TE,DOC,SS	[2]1975–83, 2005	9
43	−107.436	Elkhead Creek subwatershed	PP,DS,MI,N,TE,CB	[2]1999–2007	19
43	−107.436	Elkhead Creek subwatershed	PP,D,DS,MI,N,DOC	1999	1

Appendix 3.   Description of selected stream sampling sites in the Upper Yampa River watershed, Colorado, with type of water-quality data collected, period of water-quality record, and number of samples collected, 1975 through 2009.—Continued

[No , number; USFS, U.S. Department of the Agriculture Forest Service; PP, physical properties; D, discharge; CB, coliform bacteria; SS, suspended sediment; CDPHE, Colorado Department of Public Health and Environment; DS, dissolved solids; MI, major ions; N, nutrients, TE, trace elements; CODOW, Colorado Division of Wildlife Riverwatch Program; USGS, U.S. Geological Survey; DOC, dissolved organic carbon; CSS, City of Steamboat Springs; U, uranium; USEPA, U.S. Environmental Protection Agency. Subwatershed definitions: Yampa River subwatershed 1, Yampa River and tributaries upstream from Chuck Lewis State Wildlife Area; Yampa River subwatershed 2, Yampa River and tributaries from Chuck Lewis State Wildlife Area to Elk River confluence; Elk River subwatershed, Elk River and tributaries; Yampa River subwatershed 3, Yampa River and tributaries from Elk River confluence to Town of Hayden; Yampa River subwatershed 4, Yampa River and tributaries from Town of Hayden to Elkhead Creek confluence; Elkhead Creek subwatershed, Elkhead Creek and tributaries. Sites with the same site number are considered to be at the same location. Sites with name in bold are streamgage stations; see table 1 for additional information on these sites. The Upper Yampa River water-quality database is available at *http //rmgsc.cr.usgs.gov/cwqdr/Yampa/index.shtml*]

Site no. (see figure 3)	Site name in Upper Yampa River watershed water-quality database	Source of data	Site identifier	Latitude
44	ELKHEAD CREEK NEAR ELKHEAD, CO.	USGS	09245000	40.670
45	ENGLISH CREEK ABOVE MOUTH, NEAR CLARK, CO	USGS	404727106453700	40.791
46	FISH [Fish Creek below Hwy 40 and 10m above pedestrian bridge]	CSS	FISH	40.466
47	FISH C TRIB BL LONG LK, NR BUFFLAO PASS, CO.	USGS	09238710	40.477
48	FISH CK @ RD. 27	CDPHE	12854	40.356
49	**FISH CR AT UPPER STA NR STEAMBOAT SPRINGS, CO**	USGS	09238900	40.475
49	FISH CK AT STEAMBOAT	CDPHE	12874	40.475
50	FISH CR TRIB AB LONG LK, NR BUFFALO PASS, CO.	USGS	09238700	40.473
51	FISH CREEK AT MOUTH NEAR MILNER, CO.	USGS	402530106585700	40.425
51	FISH CREEK AT ROAD 179	CDPHE	12854A	40.423
52	FISH CREEK NEAR MILNER, CO.	USGS	09244100	40.334
53	FISH CREEK NEAR STEAMBOAT SPRINGS, CO	USGS	09239000	40.465
54	FISH CREEK NR MOUTH AT STEAMBOAT SPRINGS, CO.	USGS	402759106493100	40.466
54	FISH CK NEAR MOUTH @ HWY 40	CDPHE	12870	40.467
55	FOIDEL CREEK AT MOUTH NEAR OAK CREEK, CO	USGS	09243900	40.390
55	FOIDEL CK @ RD. 33	CDPHE	12856	40.390
56	FOIDEL CREEK NEAR OAK CREEK, CO	USGS	09243800	40.346
57	GRANITE C NR BUFFALO PASS, CO.	USGS	09238770	40.493
58	GRASSY CREEK AT GRASSY GAP, CO.	USGS	402330107082000	40.392
59	GRASSY CREEK NEAR MOUNT HARRIS, CO.	USGS	09244300	40.447
59	GRASSY CK @ RD. 27A	CDPHE	12853	40.447
60	HARRISON CREEK AT MOUTH NR BLACKTAIL MTN, CO.	USGS	402056106471600	40.349
61	HUBBERSON GULCH NEAR HAYDEN, CO.	USGS	09244464	40.391
62	L. MORRISON C AB DAM SITE NR OAK CREEK, CO	USGS	401540106502801	40.261
63	LITTLE MORRISON CK @ RD 18A	CDPHE	12896	40.273
64	LITTLE MORRISON CREEK @ CR 16	CDPHE	12896A	40.271
65	LITTLE MORRISON CREEK NEAR STAGECOACH, CO.	USGS	401634106502200	40.276
66	LITTLE WHITE SNAKE CK @ HWY 131	CDPHE	12897	40.241
67	LONG LAKE INLET NEAR BUFFALO PASS, CO.	USGS	09238705	40.474
68	LOST DOG CREEK ABOVE MOUTH, NEAR CLARK, CO	USGS	404750106454200	40.797
69	MAD CREEK NEAR STEAMBOAT SPRINGS, CO.	USGS	09242000	40.566
69	MAD CK @ CHRISTINA SWA	CDPHE	12863	40.565
70	MARTIN C AB DAM SITE NR OAK CREEK, CO	USGS	401729106514601	40.291
71	MD FK FISH C NR BUFFALO PASS, CO.	USGS	09238750	40.498
72	MID FK FISH CR TRIB BL FISH CR RESERVOIR, CO	USGS	09238800	40.497
73	MIDDLE C AB DAM SITE NR OAK CREEK, CO	USGS	401608106513001	40.269
73	MIDDLE CREEK @ CR 16	CDPHE	12809B	40.269
74	MIDDLE CK., SAMPLE #7	USGS	402354106584400	40.398
74	MIDDLE CK @ 33 RD.	CDPHE	12855	40.397
75	MIDDLE CREEK NEAR OAK CREEK, CO	USGS	09243700	40.386
76	NORTH FORK ELK RIVER ABOVE MOUTH, NEAR CLARK, CO	USGS	404620106461900	40.772
76	NORTH FORK ELK RIVER NEAR HINMAN PARK, CO.	USGS	404620106462200	40.772
77	NORTH FORK ELK RIVER ABV AGNES CREEK, NR CLARK, CO	USGS	405057106451000	40.849
78	NORTH FORK ELK RIVER ABV TRAIL CREEK, NR CLARK, CO	USGS	404950106462700	40.831
79	NORTH FORK ELKHEAD CREEK NEAR ELKHEAD, CO.	USGS	09245500	40.681
80	NORTH FORK OF FOIDEL CREEK AT MOUTH, CO	USGS	402007107050400	40.335
81	NORTH FORK WALTON CREEK NR RABBIT EARS PASS, CO.	USGS	09238300	40.396
82	OAK CR AT NAT FOREST BOUNDARY ABOVE TOWN	CDPHE	OAK01	40.217

Appendix 3.   Description of selected stream sampling sites in the Upper Yampa River watershed, Colorado, with type of water-quality data collected, period of water-quality record, and number of samples collected, 1975 through 2009.—Continued

[No , number; USFS, U.S. Department of the Agriculture Forest Service; PP, physical properties; D, discharge; CB, coliform bacteria; SS, suspended sediment; CDPHE, Colorado Department of Public Health and Environment; DS, dissolved solids; MI, major ions; N, nutrients, TE, trace elements; CODOW, Colorado Division of Wildlife Riverwatch Program; USGS, U.S. Geological Survey; DOC, dissolved organic carbon; CSS, City of Steamboat Springs; U, uranium; USEPA, U.S. Environmental Protection Agency. Subwatershed definitions: Yampa River subwatershed 1, Yampa River and tributaries upstream from Chuck Lewis State Wildlife Area; Yampa River subwatershed 2, Yampa River and tributaries from Chuck Lewis State Wildlife Area to Elk River confluence; Elk River subwatershed, Elk River and tributaries; Yampa River subwatershed 3, Yampa River and tributaries from Elk River confluence to Town of Hayden; Yampa River subwatershed 4, Yampa River and tributaries from Town of Hayden to Elkhead Creek confluence; Elkhead Creek subwatershed, Elkhead Creek and tributaries. Sites with the same site number are considered to be at the same location. Sites with name in bold are streamgage stations; see table 1 for additional information on these sites. The Upper Yampa River water-quality database is available at *http //rmgsc.cr.usgs.gov/cwqdr/Yampa/index.shtml*]

Site no. (see figure 3)	Longitude	Subwatershed	Type of water-quality data collected[1]	Period of water-quality record (calendar year)	No. of samples
44	−107.285	Elkhead Creek subwatershed	PP,D,DS,MI,N,TE,DOC,CB,SS	[3]1975–96	141
45	−106.761	Elk River subwatershed	PP,D,DS,MI,N,TE,DOC,SS	[2]1999–2003	14
46	−106.829	Yampa River subwatershed 2	PP,MI,N,TE,CB	2007–08	4
47	−106.688	Yampa River subwatershed 2	PP,D,DS,MI,N,SS	1985–95	40
48	−107.104	Yampa River subwatershed 3	PP,DS,MI,N,TE,CB	2001	4
49	−106.787	Yampa River subwatershed 2	PP,D,DS,MI,N,TE,CB,SS	1982–2004	166
49	−106.783	Yampa River subwatershed 2	PP,DS,MI,N,TE	1996–97	10
50	−106.680	Yampa River subwatershed 2	PP,D	1985–87	8
51	−106.983	Yampa River subwatershed 3	PP,D,DS,MI,N,TE,DOC,SS	[2]1975–82, 2005	16
51	−106.986	Yampa River subwatershed 3	PP,MI,N,TE	2006–07	4
52	−107.139	Yampa River subwatershed 3	PP,D,DS,MI,N,TE,DOC,SS	[2]1975–82, 2005	18
53	−106.821	Yampa River subwatershed 2	PP,D,N,CB,SS	1975–76	8
54	−106.826	Yampa River subwatershed 2	PP,D,N,CB	1975, 2005	4
54	−106.825	Yampa River subwatershed 2	PP,D,DS,MI,N,TE,CB	[2]1999–2002	5
55	−106.995	Yampa River subwatershed 3	PP,D,DS,MI,N,TE,DOC,SS	1975–2001, 2005	290
55	−106.996	Yampa River subwatershed 3	PP,DS,MI,N,TE,CB	[2]2001–07	7
56	−107.085	Yampa River subwatershed 3	PP,D,DS,MI,N,TE,DOC,SS	[4]1975–2001	226
57	−106.693	Yampa River subwatershed 2	PP,D,MI,N,SS	1985–95	65
58	−107.139	Yampa River subwatershed 3	PP,D,DS,MI,N,TE,DOC,SS	[2]1975–82	22
59	−107.146	Yampa River subwatershed 3	PP,D,DS,MI,N,TE,DOC,SS	[2]1975–82	21
59	−107.146	Yampa River subwatershed 3	PP,DS,MI,N,TE,U,CB	[2]2001–07	7
60	−106.788	Yampa River subwatershed 1	PP,D,CB,SS	1975–76	4
61	−107.271	Yampa River subwatershed 4	PP,D,DS,MI,N,TE,DOC,SS	1978–82	41
62	−106.842	Yampa River subwatershed 1	PP,D,SS	1986–88	17
63	−106.839	Yampa River subwatershed 1	PP,DS,MI,N,TE	[2]2001–07	7
64	−106.839	Yampa River subwatershed 1	PP,D,DS,MI,N,TE	2000	1
65	−106.840	Yampa River subwatershed 1	PP,D,N,CB,SS	1975–76	5
66	−106.943	Yampa River subwatershed 1	PP,DS,MI,N,TE	[2]2001–07	4
67	−106.680	Yampa River subwatershed 2	PP,D,MI,N,SS	1986–95	51
68	−106.762	Elk River subwatershed	PP,D,DS,MI,N,TE,U,DOC,SS	[2]1999–2006	21
69	−106.889	Elk River subwatershed	PP,D,DS,MI,N,TE,DOC,CB,SS	1975–76, 2003	12
69	−106.889	Elk River subwatershed	PP,DS,MI,N,TE	1996–98, 2006–07	16
70	−106.863	Yampa River subwatershed 1	PP,D,SS	1986–88	17
71	−106.692	Yampa River subwatershed 2	PP,D,MI,N,SS	1985–95	61
72	−106.699	Yampa River subwatershed 2	PP,D,MI,N,SS	[2]1985–94	15
73	−106.859	Yampa River subwatershed 1	PP,D,SS	1986–88	16
73	−106.859	Yampa River subwatershed 1	PP,D,DS,MI,N,TE	2000	1
74	−106.979	Yampa River subwatershed 3	PP,D	2005	1
74	−106.980	Yampa River subwatershed 3	PP,DS,MI,N,TE,CB	2001, 2006–07	8
75	−106.993	Yampa River subwatershed 3	PP,D,DS,MI,N,TE,DOC,SS	1975–2001	243
76	−106.773	Elk River subwatershed	PP,D,DS,MI,N,TE,U,DOC,SS	[2]1999–2006	23
76	−106.773	Elk River subwatershed	PP,D,CB,SS	1975–76	4
77	−106.753	Elk River subwatershed	PP,D,DS,MI,N,TE,SS	1999–2000	11
78	−106.775	Elk River subwatershed	PP,D,DS,MI,N,TE,SS	1999–2000	11
79	−107.287	Elkhead Creek subwatershed	PP,D,CB,SS	1975–76	4
80	−107.085	Yampa River subwatershed 3	PP,D,DS,MI,N	1983	2
81	−106.650	Yampa River subwatershed 2	PP,D,DS,MI,N,TE,DOC	[2]1975–87	10
82	−107.067	Yampa River subwatershed 1	PP,DS,TE	1979	1

Appendix 3.   Description of selected stream sampling sites in the Upper Yampa River watershed, Colorado, with type of water-quality data collected, period of water-quality record, and number of samples collected, 1975 through 2009.—Continued

[No., number; USFS, U.S. Department of the Agriculture Forest Service; PP, physical properties; D, discharge; CB, coliform bacteria; SS, suspended sediment; CDPHE, Colorado Department of Public Health and Environment; DS, dissolved solids; MI, major ions; N, nutrients, TE, trace elements; CODOW, Colorado Division of Wildlife Riverwatch Program; USGS, U.S. Geological Survey; DOC, dissolved organic carbon; CSS, City of Steamboat Springs; U, uranium; USEPA, U.S. Environmental Protection Agency. Subwatershed definitions: Yampa River subwatershed 1, Yampa River and tributaries upstream from Chuck Lewis State Wildlife Area; Yampa River subwatershed 2, Yampa River and tributaries from Chuck Lewis State Wildlife Area to Elk River confluence; Elk River subwatershed, Elk River and tributaries; Yampa River subwatershed 3, Yampa River and tributaries from Elk River confluence to Town of Hayden; Yampa River subwatershed 4, Yampa River and tributaries from Town of Hayden to Elkhead Creek confluence; Elkhead Creek subwatershed, Elkhead Creek and tributaries. Sites with the same site number are considered to be at the same location. Sites with name in bold are streamgage stations; see table 1 for additional information on these sites. The Upper Yampa River water-quality database is available at *http //rmgsc.cr.usgs.gov/cwqdr/Yampa/index.shtml*]

Site no. (see figure 3)	Site name in Upper Yampa River watershed water-quality database	Source of data	Site identifier	Latitude
83	OAK CK D/S TOWN OF OAK CREEK @ CR 27	CDPHE	12892	40.276
84	OAK CK NEAR MOUTH @ CR 14 @ SYDNEY PEAK HORSE RANCH	CDPHE	12891A	40.390
85	OAK CREEK AB OAK CREEK DRAIN NEAR OAK CREEK, CO.	USGS	401725106575600	40.290
86	OAK CREEK ABOVE ROUTT, CO.	USGS	401741106574600	40.295
87	OAK CREEK AT CR 35 BELOW HAYBRO, CO	USGS	402121106543201	40.356
88	OAK CREEK AT P AND M COAL MINE OAK CREEK COLO	CDPHE	OAK02	40.3
89	OAK CREEK AT WHITECOTTON RD	CDPHE	OAK03	40.333
90	OAK CREEK BELOW TOWN OF OAK CREEK	CDPHE	000153	40.332
91	Oak Creek Decker Pk	CODOW	CDOWRW-80	40.27
92	Oak Creek Habro Br	CODOW	CDOWRW-9	40.31
93	OAK CREEK NEAR OAK CREEK, CO.	USGS	09238000	40.244
94	OAK CREEK NEAR STEAMBOAT SPRINGS, CO.	USGS	402356106503000	40.399
94	OAK CK @ 22 RD ABV YAMPA R	CDPHE	12891	40.399
95	PHILLIPS CREEK NEAR YAMPA, CO.	USGS	400759106532500	40.133
96	PRIEST C BL SKI AREA NR STEAMBOAT SPRINGS, CO.	USGS	402600106473600	40.433
97	S. FORK ELK R. @ TRAIL 1169 NR FR 443	CDPHE	12869	40.751
98	SAGE CK @ RD. 27	CDPHE	12851	40.484
99	SAGE CR AB HADEN STATION	USGS	402855107101501	40.482
100	SAGE CREEK ABOVE SAGE CREEK RES, NR HAYDEN, CO.	USGS	09244415	40.384
101	SAGE CREEK NEAR HAYDEN, CO.	USGS	402918107094400	40.488
101	SAGE CR @ HWY 40	USGS	402917107094501	40.488
102	SAGE CREEK NEAR MOUNT HARRIS, CO.	USGS	402522107134100	40.423
103	SCC87-CREEK	USGS	402358107054601	40.399
104	SENECA NORTH	USGS	402940107074200	40.494
105	SENECA NORTHEAST	USGS	402915107074500	40.487
106	SMUIN GULCH NEAR HAYDEN, CO	USGS	402829107193700	40.475
107	SMUIN TRIB. CREEK NEAR HAYDEN, CO	USGS	402845107185100	40.479
108	SODA CREEK NR MOUTH AT STEAMBOAT SPRINGS,CO.	USGS	402920106501900	40.489
108	SODA [Soda Creek abv second pedestrian bridge from confl with Yampa R in Little Toots Park	CSS	SODA	40.488
109	SOUTH FORK OF FOIDEL CREEK ABOVE MINE, CO	USGS	401847107063400	40.313
110	SOUTH FORK OF FOIDEL CREEK AT MOUTH, CO	USGS	402008107050200	40.336
111	SPRING CREEK NR MOUTH AT STEAMBOAT SPRINGS, CO.	USGS	402857106494000	40.482
112	STOKES GULCH NEAR HAYDEN, CO.	USGS	09244470	40.468
113	TOW CREEK NEAR MOUTH AT US 40 BRIDGE, CO.	USGS	402908107025100	40.486
114	TROUT CK @ RD. 27	CDPHE	12876H	40.312
115	TROUT CK NR. MOUTH	CDPHE	12876	40.460
116	TROUT CREEK ABOVE FOIDEL CREEK NEAR MILNER, CO.	USGS	402416106580800	40.404
117	TROUT CREEK ABOVE MILNER, CO.	USGS	402720106591200	40.456
118	TROUT CREEK BELOW FOIDEL CREEK NEAR MILNER, CO.	USGS	402536106582700	40.427
119	TROUT CREEK NEAR MILNER, CO.	USGS	402816107003800	40.471
120	TROUT CREEK NEAR OAK CREEK, CO	USGS	401816107011000	40.304
121	TROUT CREEK NEAR PHIPPSBURG, CO.	USGS	09243000	40.151
122	WALTON CR. NEAR MOUTH @ HWY 40	CDPHE	12894	40.270
123	**WALTON CREEK NEAR STEAMBOAT SPRINGS, CO.**[6]	USGS	09238500	40.408
124	WALTON CREEK NR MOUTH AT US 40 BRIDGE, CO.	USGS	402700106485400	40.450
124	WALTON CREEK @ HWY 40	CDPHE	12894A	40.450
124	WALT [Walton Creek 10m above Hwy 40 bridge]	CSS	WALT	40.45

Appendix 3. Description of selected stream sampling sites in the Upper Yampa River watershed, Colorado, with type of water-quality data collected, period of water-quality record, and number of samples collected, 1975 through 2009.—Continued

[No , number; USFS, U.S. Department of the Agriculture Forest Service; PP, physical properties; D, discharge; CB, coliform bacteria; SS, suspended sediment; CDPHE, Colorado Department of Public Health and Environment; DS, dissolved solids; MI, major ions; N, nutrients, TE, trace elements; CODOW, Colorado Division of Wildlife Riverwatch Program; USGS, U.S. Geological Survey; DOC, dissolved organic carbon; CSS, City of Steamboat Springs; U, uranium; USEPA, U.S. Environmental Protection Agency. Subwatershed definitions: Yampa River subwatershed 1, Yampa River and tributaries upstream from Chuck Lewis State Wildlife Area; Yampa River subwatershed 2, Yampa River and tributaries from Chuck Lewis State Wildlife Area to Elk River confluence; Elk River subwatershed, Elk River and tributaries; Yampa River subwatershed 3, Yampa River and tributaries from Elk River confluence to Town of Hayden; Yampa River subwatershed 4, Yampa River and tributaries from Town of Hayden to Elkhead Creek confluence; Elkhead Creek subwatershed, Elkhead Creek and tributaries. Sites with the same site number are considered to be at the same location. Sites with name in bold are streamgage stations; see table 1 for additional information on these sites. The Upper Yampa River water-quality database is available at *http //rmgsc.cr.usgs.gov/cwqdr/Yampa/index.shtml*]

Site no. (see figure 3)	Longitude	Subwatershed	Type of water-quality data collected[1]	Period of water-quality record (calendar year)	No. of samples
83	−106.964	Yampa River subwatershed 1	PP,D,DS,MI,N,TE,CB	[2]1999–2007	14
84	−106.843	Yampa River subwatershed 1	PP,MI,N,TE	2006–07	4
85	−106.966	Yampa River subwatershed 1	PP,D,N,TE,DOC,CB,SS	1975–76, 2005	7
86	−106.963	Yampa River subwatershed 1	PP,D,DS,MI,N,TE,DOC,SS	1975–76	5
87	−106.909	Yampa River subwatershed 1	PP,D	2005	2
88	−107.033	Yampa River subwatershed 1	PP,DS,TE	1979	1
89	−106.967	Yampa River subwatershed 1	PP,DS,TE	1979	1
90	−106.96	Yampa River subwatershed 1	PP,DS,MI,N,TE,U	1979–92	74
91	−106.95	Yampa River subwatershed 1	PP,TE	[5]1991–2000	92
92	−106.95	Yampa River subwatershed 1	PP,TE	[5]1991–99	110
93	−107.015	Yampa River subwatershed 1	PP,D,DS,MI,N,TE,DOC,SS	[2]1975–81, 2005	13
94	−106.842	Yampa River subwatershed 1	PP,D,N,TE,DOC	1975, 2005	2
94	−106.842	Yampa River subwatershed 1	PP,D,DS,N,TE,U,CB	[2]1996–2004	25
95	−106.891	Yampa River subwatershed 1	PP,D,TE	1975, 2005	2
96	−106.794	Yampa River subwatershed 2	PP,D,N	1976	3
97	−106.731	Elk River subwatershed	PP,DS,MI,N,TE,CB	2001	6
98	−107.170	Yampa River subwatershed 3	PP,DS,MI,N,TE,U,CB	2001, 2004–07	7
99	−107.171	Yampa River subwatershed 3	PP,D,DS,MI,N,TE	1979	2
100	−107.193	Yampa River subwatershed 3	PP,D,DS,MI,N,TE,DOC,SS	1981–83	28
101	−107.163	Yampa River subwatershed 3	PP,D,MI,N,TE,DOC,CB,SS	1975–76, 2005	20
101	−107.163	Yampa River subwatershed 3	PP,D,DS,MI,N,TE	1979	3
102	−107.229	Yampa River subwatershed 3	PP,D,MI,N,TE,DOC	1975–76	5
103	−107.097	Yampa River subwatershed 3	PP,DS,MI,N,TE	1988	3
104	−107.129	Yampa River subwatershed 3	PP,DS,MI,N,TE	1986	1
105	−107.130	Yampa River subwatershed 3	PP,DS,MI,N,TE	1986	1
106	−107.328	Yampa River subwatershed 4	PP,D,DS,MI,N,TE,SS	1981–82	6
107	−107.315	Yampa River subwatershed 4	PP,D,DS,MI,N	1981	6
108	−106.839	Yampa River subwatershed 2	PP,D,N,CB	1975–76, 2005	6
108	−106.84	Yampa River subwatershed 2	PP,MI,N,TE,CB	2007–08	4
109	−107.110	Yampa River subwatershed 3	PP,D,DS,MI,N	1983	2
110	−107.084	Yampa River subwatershed 3	PP,D,DS,MI,N	1983	1
111	−106.828	Yampa River subwatershed 2	PP,D,N,CB	1975–76, 2005	6
112	−107.247	Yampa River subwatershed 4	PP,D,DS,MI,N,TE,DOC,SS	1978–82	39
113	−107.048	Yampa River subwatershed 3	D,CB	1975	1
114	−107.009	Yampa River subwatershed 3	PP,DS,MI,N,TE	2001	4
115	−106.989	Yampa River subwatershed 3	PP,DS,MI,N,TE,U	[2]1996–2007	22
116	−106.969	Yampa River subwatershed 3	PP,D,N,TE,DOC	1975, 2005	2
117	−106.987	Yampa River subwatershed 3	PP,D,DS,MI,N,TE,SS	1981–82	11
118	−106.975	Yampa River subwatershed 3	PP,D,DS,MI,N,TE,DOC,SS	1975, 2005	2
119	−107.011	Yampa River subwatershed 3	PP,D,DS,MI,N,DOC,CB	1975, 1979, 2005	6
120	−107.020	Yampa River subwatershed 3	PP,D,DS,MI,N	1981	7
121	−107.132	Yampa River subwatershed 3	PP,D,DS,MI,N,TE,DOC	1975	1
122	−106.816	Yampa River subwatershed 2	PP,D,DS,MI,N,TE,CB	1999, 2006–07	5
123	−106.787	Yampa River subwatershed 2	PP,D,DS,MI,N,TE,SS	1982–87	45
124	−106.816	Yampa River subwatershed 2	PP,D,N,CB	1975, 2005	4
124	−106.816	Yampa River subwatershed 2	PP,DS,MI,N,TE,CB	2001–02	3
124	−106.815	Yampa River subwatershed 2	PP,MI,N,TE,CB	2007–08	4

Appendix 3. Description of selected stream sampling sites in the Upper Yampa River watershed, Colorado, with type of water-quality data collected, period of water-quality record, and number of samples collected, 1975 through 2009.—Continued

[No , number; USFS, U.S. Department of the Agriculture Forest Service; PP, physical properties; D, discharge; CB, coliform bacteria; SS, suspended sediment; CDPHE, Colorado Department of Public Health and Environment; DS, dissolved solids; MI, major ions; N, nutrients, TE, trace elements; CODOW, Colorado Division of Wildlife Riverwatch Program; USGS, U.S. Geological Survey; DOC, dissolved organic carbon; CSS, City of Steamboat Springs; U, uranium; USEPA, U.S. Environmental Protection Agency. Subwatershed definitions: Yampa River subwatershed 1, Yampa River and tributaries upstream from Chuck Lewis State Wildlife Area; Yampa River subwatershed 2, Yampa River and tributaries from Chuck Lewis State Wildlife Area to Elk River confluence; Elk River subwatershed, Elk River and tributaries; Yampa River subwatershed 3, Yampa River and tributaries from Elk River confluence to Town of Hayden; Yampa River subwatershed 4, Yampa River and tributaries from Town of Hayden to Elkhead Creek confluence; Elkhead Creek subwatershed, Elkhead Creek and tributaries. Sites with the same site number are considered to be at the same location. Sites with name in bold are streamgage stations; see table 1 for additional information on these sites. The Upper Yampa River water-quality database is available at *http //rmgsc.cr.usgs.gov/cwqdr/Yampa/index.shtml*]

Site no. (see figure 3)	Site name in Upper Yampa River watershed water-quality database	Source of data	Site identifier	Latitude
125	WATERING TROUGH GULCH NEAR HAYDEN, CO.	USGS	09244460	40.382
126	WAYS GULCH AT HAHNS PEAK, CO SITE 1A	USGS	404756106555100	40.799
127	WEST FORK ELK RIVER NR MOUTH AT US 40 BRIDGE,CO.	USGS	402903106584100	40.484
128	WOLF CREEK NEAR HAYDEN, CO.	USGS	402832107080200	40.476
128	WOLF CREEK AT HWY 40	CDPHE	12802C	40.476
129	YAMPA ABOVE OAK CREEK CONFLUENCE	CDPHE	000088	40.383
130	YAMPA R 0.5 MI DSTRM STP BELOW STMBT SPRGS.	USGS	403015106523000	40.504
131	YAMPA R AB DAM SITE NR OAK CREEK, CO	USGS	401609106525201	40.269
132	YAMPA R AT 13TH ST BRIDGE AT STEAMBOAT SPRINGS, CO	USGS	402922106502701	40.489
133	YAMPA R AT JAMES BRN BR BLW STEAMBOAT SPRINGS, CO	USGS	402946106512601	40.496
133	YMP-7 [Yampa River 100m above James Brown Bridge]	CSS	YMP-7	40.496
134	YAMPA R AT US HWY 40	CDPHE	12802B	40.495
135	YAMPA R BL KOA CAMPGROUNDS NR STEAMBOAT SPG, CO.	USGS	403017106525800	40.505
136	YAMPA R. @ CR 14 FISHING ACCESS	CDPHE	12806D	40.475
137	YAMPA R. ABV. PHIPPSBURG	CDPHE	12814	40.227
138	YAMPA R. BLW STAGECOACH RES.	CDPHE	12808	40.287
139	YAMPA R. D/S STAGECOACH RES. DAM	CDPHE	12808P	40.288
140	YAMPA R. N. OF HAYDEN @ CALIFORNIA PARK RD	CDPHE	12802	40.502
141	YAMPA R. NR MOUNT HARRIS BLW HWY 40 BRIDGE	CDPHE	12805	40.488
142	**YAMPA R. U/S LAKE CATAMOUNT @ CR18**[7]	CDPHE	12807	40.341
143	YAMPA RIVER AB OAK CREEK NR STEAMBOAT SPGS, CO.	USGS	402356106500000	40.399
143	YAMPA R. ABV OAK CREEK	CDPHE	12811	40.399
144	YAMPA RIVER AB SEWAGE PLANT BL STEAMBOAT SPG,CO.	USGS	402934106505400	40.493
145	YAMPA RIVER ABOVE ELK RIVER NEAR MILNER, CO.	USGS	402932106564900	40.492
145	YAMPA RIVER ABV ELK RIVER	USGS	402936106565000	40.493
146	**YAMPA RIVER ABOVE ELKHEAD CREEK NEAR HAYDEN, CO**	USGS	09244490	40.518
147	**YAMPA RIVER ABOVE STAGECOACH RESERVOIR, CO**	USGS	09237450	40.269
147	YAMPA R. U/S STAGECOACH RES @ CR16	CDPHE	12809	40.269
148	YAMPA RIVER ABOVE TOW CREEK OIL FIELD, CO.	USGS	402902107043600	40.484
149	YAMPA RIVER AT ELK RIVER JUNCTION NR MILNER, CO.	USGS	402902106580000	40.484
150	YAMPA RIVER AT HAYDEN	USGS	403007107155001	40.502
150	YAMPA RIVER AT HAYDEN, CO.	USGS	403006107154800	40.502
151	YAMPA RIVER AT MILNER, CO.	USGS	402840107004200	40.478
151	YAMPA RIVER AT MILNER	CDPHE	000038	40.479
152	YAMPA RIVER AT PHIPPSBURG, CO.	USGS	401418106562200	40.238
153	**YAMPA RIVER AT STEAMBOAT SPRINGS, CO**	USGS	09239500	40.483
153	YAMPA R. @ 5TH ST. BRIDGE IN STEAMBOAT	CDPHE	12806	40.483
153	Yampa River 5th St Br	CODOW	CDOWRW-607	40.48
154	YAMPA RIVER BELOW DIVERSION, NEAR HAYDEN, CO.	USGS	09244410	40.488
155	YAMPA RIVER BELOW HAYDEN, CO.	USGS	402930107174200	40.492
155	YAMPA R. WEST OF HAYDEN @ HWY 40	CDPHE	12802A	40.492
156	YAMPA RIVER BELOW MORGAN CREEK NEAR HAYDEN, CO.	USGS	403051107124500	40.514
157	YAMPA RIVER BELOW OAK CREEK NR STEAMBOAT SPG,CO.	USGS	402544106493600	40.429
158	Yampa River Below Stagecoach	CODOW	CDOWRW-81	40.28
158	**YAMPA RIVER BELOW STAGECOACH RESERVOIR, CO**	USGS	09237500	40.285
159	YAMPA RIVER BELOW STEAMBOAT II SEWAGE PLANT, CO.	USGS	403002106545500	40.501
160	YAMPA RIVER BELOW TROUT CREEK AT MILNER, CO.	USGS	402854107020500	40.482
161	YAMPA RIVER BELOW WALTON CREEK, CO.	USGS	402737106493700	40.460

Appendix 3.   Description of selected stream sampling sites in the Upper Yampa River watershed, Colorado, with type of water-quality data collected, period of water-quality record, and number of samples collected, 1975 through 2009.—Continued

[No , number; USFS, U.S. Department of the Agriculture Forest Service; PP, physical properties; D, discharge; CB, coliform bacteria; SS, suspended sediment; CDPHE, Colorado Department of Public Health and Environment; DS, dissolved solids; MI, major ions; N, nutrients; TE, trace elements; CODOW, Colorado Division of Wildlife Riverwatch Program; USGS, U.S. Geological Survey; DOC, dissolved organic carbon; CSS, City of Steamboat Springs; U, uranium; USEPA, U.S. Environmental Protection Agency. Subwatershed definitions: Yampa River subwatershed 1, Yampa River and tributaries upstream from Chuck Lewis State Wildlife Area; Yampa River subwatershed 2, Yampa River and tributaries from Chuck Lewis State Wildlife Area to Elk River confluence; Elk River subwatershed, Elk River and tributaries; Yampa River subwatershed 3, Yampa River and tributaries from Elk River confluence to Town of Hayden; Yampa River subwatershed 4, Yampa River and tributaries from Town of Hayden to Elkhead Creek confluence; Elkhead Creek subwatershed, Elkhead Creek and tributaries. Sites with the same site number are considered to be at the same location. Sites with name in bold are streamgage stations; see table 1 for additional information on these sites. The Upper Yampa River water-quality database is available at *http //rmgsc.cr.usgs.gov/cwqdr/Yampa/index.shtml*]

Site no. (see figure 3)	Longitude	Subwatershed	Type of water-quality data collected[1]	Period of water-quality record (calendar year)	No. of samples
125	−107.281	Yampa River subwatershed 4	PP,D,DS,MI,N,TE,DOC,SS	1978–81	27
126	−106.931	Elk River subwatershed	PP,D,MI,N,TE,DOC	1975	1
127	−106.979	Yampa River subwatershed 3	PP,D,N,CB	1975, 2005	5
128	−107.134	Yampa River subwatershed 3	PP,D,N,TE,DOC	1975	3
128	−107.134	Yampa River subwatershed 3	PP,MI,N,TE	2006–07	4
129	−106.817	Yampa River subwatershed 1	PP,DS,MI,N,TE,U	1975-93	101
130	−106.876	Yampa River subwatershed 2	PP,N,CB	1976	2
131	−106.882	Yampa River subwatershed 1	PP,D,SS	1986–88	17
132	−106.841	Yampa River subwatershed 2	PP	2005	1
133	−106.857	Yampa River subwatershed 2	PP	2005	1
133	−106.857	Yampa River subwatershed 2	PP,MI,N,TE,CB	2007–08	4
134	−107.158	Yampa River subwatershed 3	PP,MI,N,TE	2006–07	4
135	−106.883	Yampa River subwatershed 2	PP,D,N,CB	1975–76	5
136	−106.824	Yampa River subwatershed 1	PP,DS,MI,N,TE	1999	1
137	−106.941	Yampa River subwatershed 1	PP,D,DS,MI,N,TE,U,CB	1996–2004	34
138	−106.829	Yampa River subwatershed 1	PP,D,DS,MI,N,TE	1996–99	14
139	−106.827	Yampa River subwatershed 1	PP,DS,MI,N,TE,CB	[2]2000–07	21
140	−107.264	Yampa River subwatershed 3	PP,DS,MI,N,TE,CB	[1]1996–99	13
141	−107.158	Yampa River subwatershed 3	PP,DS,MI,N,TE	1999	1
142	−106.808	Yampa River subwatershed 1	PP,DS,MI,N,TE,CB	1999	1
143	−106.834	Yampa River subwatershed 1	PP,D,DS,MI,N,TE,DOC,CB,SS	1975–76, 2005	6
143	−106.834	Yampa River subwatershed 1	PP,D,DS,MI,N,TE,CB	1996–2002	30
144	−106.849	Yampa River subwatershed 2	PP,D,N,CB,SS	1975–76, 2005	11
145	−106.948	Yampa River subwatershed 2	PP,D,N,TE,CB,DOC	1975–76	10
145	−106.948	Yampa River subwatershed 2	PP,D,DS,MI,N,DOC	[2]1999–2005	5
146	−107.400	Yampa River subwatershed 4	PP,D	2004–05	13
147	−106.881	Yampa River subwatershed 1	PP,D,DS,MI,N,TE,CB,SS	1988–2005	202
147	−106.881	Yampa River subwatershed 1	PP,DS,MI,N,TE,CB	[2]1999–2007	22
148	−107.077	Yampa River subwatershed 3	PP,D,N,CB	1975	3
149	−106.967	Yampa River subwatershed 2	PP,D,N,CB	1975–76	5
150	−107.265	Yampa River subwatershed 3	PP,D,TE	1979	1
150	−107.264	Yampa River subwatershed 3	PP,D,MI,N,TE,DOC,CB	1975–76, 2005	10
151	−107.012	Yampa River subwatershed 3	PP,D,N,CB	1975, 2005	5
151	−107.013	Yampa River subwatershed 3	PP,DS,MI,N,TE,U,CB	[8]1975–2007	179
152	−106.940	Yampa River subwatershed 1	PP,D,N,TE,DOC,CB,SS	1975–76, 2005	7
153	−106.832	Yampa River subwatershed 2	PP,D,DS,MI,N,TE,DOC,CB,SS	1975–2009	358
153	−106.832	Yampa River subwatershed 2	PP,DS,MI,N,TE,U,CB	[8]1996–2007	38
153	−106.83	Yampa River subwatershed 2	PP,D,TE	1998–2000	17
154	−107.160	Yampa River subwatershed 3	PP,D,DS,MI,N,TE,DOC,CB,SS	[2]1975–2005	158
155	−107.296	Yampa River subwatershed 4	PP,D,DS,MI,N,TE,DOC,CB	1975–76, 2005	8
155	−107.296	Yampa River subwatershed 4	PP,DS,MI,N,TE,CB	2000–02	18
156	−107.213	Yampa River subwatershed 3	PP,D,N,CB	1975	3
157	−106.827	Yampa River subwatershed 2	PP,D,N,TE,DOC,CB,SS	1975–76, 2005	10
158	−106.82	Yampa River subwatershed 1	PP,D,MI,N,TE	[9]1991–2004	134
158	−106.831	Yampa River subwatershed 1	PP,D,DS,MI,N,TE,DOC,CB,SS	1984–2005	263
159	−106.916	Yampa River subwatershed 2	PP,D,N,CB	1975–76	5
160	−107.035	Yampa River subwatershed 3	PP,D,N,TE,DOC,CB	1975–76, 2005	9
161	−106.828	Yampa River subwatershed 2	PP,N,CB	1975–76	6

Appendix 3. Description of selected stream sampling sites in the Upper Yampa River watershed, Colorado, with type of water-quality data collected, period of water-quality record, and number of samples collected, 1975 through 2009.—Continued

[No., number; USFS, U.S. Department of the Agriculture Forest Service; PP, physical properties; D, discharge; CB, coliform bacteria; SS, suspended sediment; CDPHE, Colorado Department of Public Health and Environment; DS, dissolved solids; MI, major ions; N, nutrients, TE, trace elements; CODOW, Colorado Division of Wildlife Riverwatch Program; USGS, U.S. Geological Survey; DOC, dissolved organic carbon; CSS, City of Steamboat Springs; U, uranium; USEPA, U.S. Environmental Protection Agency. Subwatershed definitions: Yampa River subwatershed 1, Yampa River and tributaries upstream from Chuck Lewis State Wildlife Area; Yampa River subwatershed 2, Yampa River and tributaries from Chuck Lewis State Wildlife Area to Elk River confluence; Elk River subwatershed, Elk River and tributaries; Yampa River subwatershed 3, Yampa River and tributaries from Elk River confluence to Town of Hayden; Yampa River subwatershed 4, Yampa River and tributaries from Town of Hayden to Elkhead Creek confluence; Elkhead Creek subwatershed, Elkhead Creek and tributaries. Sites with the same site number are considered to be at the same location. Sites with name in bold are streamgage stations; see table 1 for additional information on these sites. The Upper Yampa River water-quality database is available at *http //rmgsc.cr.usgs.gov/cwqdr/Yampa/index.shtml*]

Site no. (see figure 3)	Site name in Upper Yampa River watershed water-quality database	Source of data	Site identifier	Latitude
162	YAMPA RIVER BELOW YAMPA, CO.	USGS	401048106544800	40.180
162	Yampa River Willow Cabin Br	CODOW	CDOWRW-7	40.18
162	YAMPA R. BLW YAMPA @ CR21	CDPHE	12815	40.183
163	YAMPA RIVER BL SEWAGE PLANT BL STEAMBOAT SPG,CO.	USGS	402958106515200	40.499
163	YAMPA R 0.2 MI BL STP BL STEAMBOAT SPRINGS, CO.	USGS	402958106515201	40.499
164	Yampa River East Br	CODOW	CDOWRW-13	40.48
165	Yampa River Hunt Creek	CODOW	CDOWRW-3260	40.22
166	Yampa River Library	CODOW	CDOWRW-12	40.48
167	Yampa River N of Stagecoach	CODOW	CDOWRW-3476	40.29
168	YAMPA RIVER NEAR HAYDEN, CO.	USGS	09244400	40.489
169	YAMPA RIVER NR SIDNEY, COLO.	USGS	402230106493000	40.375
169	YAMPA R. D/S LAKE CATAMOUNT @ HWY 131	CDPHE	12806F	40.375
170	Yampa River Stagecoach Res	CODOW	CDOWRW-8	40.26
171	Yampa River SWA Br	CODOW	CDOWRW-10	40.39
172	Yampa River Treehouse	CODOW	CDOWRW-11	40.48
173	Yampa River West Br	CODOW	CDOWRW-14	40.49
174	YMP-1 [Yampa River 200m above confl Walton Creek]	CSS	YMP-1	40.449
175	YMP-2 [Yampa River 35m above confl Fish Creek]	CSS	YMP-2	40.466
176	YMP-3A [Yampa River 70m above pedestrian bridge and hot spring outflow in Weiss Park]	CSS	YMP-3A	40.481

Appendix 3. Description of selected stream sampling sites in the Upper Yampa River watershed, Colorado, with type of water-quality data collected, period of water-quality record, and number of samples collected, 1975 through 2009.—Continued

[No., number; USFS, U.S. Department of the Agriculture Forest Service; PP, physical properties; D, discharge; CB, coliform bacteria; SS, suspended sediment; CDPHE, Colorado Department of Public Health and Environment; DS, dissolved solids; MI, major ions; N, nutrients, TE, trace elements; CODOW, Colorado Division of Wildlife Riverwatch Program; USGS, U.S. Geological Survey; DOC, dissolved organic carbon; CSS, City of Steamboat Springs; U, uranium; USEPA, U.S. Environmental Protection Agency. Subwatershed definitions: Yampa River subwatershed 1, Yampa River and tributaries upstream from Chuck Lewis State Wildlife Area; Yampa River subwatershed 2, Yampa River and tributaries from Chuck Lewis State Wildlife Area to Elk River confluence; Elk River subwatershed, Elk River and tributaries; Yampa River subwatershed 3, Yampa River and tributaries from Elk River confluence to Town of Hayden; Yampa River subwatershed 4, Yampa River and tributaries from Town of Hayden to Elkhead Creek confluence; Elkhead Creek subwatershed, Elkhead Creek and tributaries. Sites with the same site number are considered to be at the same location. Sites with name in bold are streamgage stations; see table 1 for additional information on these sites. The Upper Yampa River water-quality database is available at *http //rmgsc.cr.usgs.gov/cwqdr/Yampa/index.shtml*]

Site no. (see figure 3)	Longitude	Subwatershed	Type of water-quality data collected[1]	Period of water-quality record (calendar year)	No. of samples
162	−106.914	Yampa River subwatershed 1	PP,D,TE	1975, 2005	3
162	−106.91	Yampa River subwatershed 1	PP,D	1990–99	131
162	−106.915	Yampa River subwatershed 1	PP,D,DS,MI,N,TE,CB	1999–2002	14
163	−106.865	Yampa River subwatershed 2	PP,D,N,CB	1975	3
163	−106.865	Yampa River subwatershed 2	PP,D,N,CB	1976	2
164	−107.15	Yampa River subwatershed 3	PP,D,TE	1990–96, 1998	49
165	−106.94	Yampa River subwatershed 1	PP,TE	1993	1
166	−106.84	Yampa River subwatershed 2	PP,TE	1991–92, 1996	23
167	−106.8	Yampa River subwatershed 1	PP	1993	1
168	−107.160	Yampa River subwatershed 3	PP,D,DS,MI,N,TE	1975, 1979, 2005	6
169	−106.826	Yampa River subwatershed 1	PP	2005	1
169	−106.825	Yampa River subwatershed 1	PP,DS,MI,N,TE,CB	1999	1
170	−106.88	Yampa River subwatershed 1	PP,D,MI,N,TE	[9]1990–2004	150
171	−106.83	Yampa River subwatershed 1	PP,D,TE	1991–98	147
172	−106.83	Yampa River subwatershed 2	PP,D,TE	1991, 1996	15
173	−107.29	Yampa River subwatershed 4	PP,TE	[2]1990–2001	45
174	−106.820	Yampa River subwatershed 2	PP,MI,N,TE,CB	2007–08	4
175	−106.830	Yampa River subwatershed 2	PP,MI,N,TE,CB	2007–08	4
176	−106.828	Yampa River subwatershed 2	PP,MI,N,TE,CB	2007–08	4

[1]When multiple sample were collected at a site, all types of water-quality data listed may not be available for each sample.

[2]Samples were not collected every year in the period of record.

[3]Samples were not collected in 1978.

[4]Samples were not collected in 1984.

[5]Samples were not collected in 1997.

[6]Site also is Colorado Division of Water Resources streamflow gage WLTNCKCO. See table 1 for additional information.

[7]Site also is Colorado Division of Water Resources streamflow gage YAMABVCO. See table 1 for additional information.

[8]Samples were not collected in 2005.

[9]Samples were not collected in 2001.

Appendix 4.   Description of selected lake and reservoir samlping sites in the Upper Yampa River watershed, Colorado, with type of water-quality data collected, period of water-quality record, and number of sample days, 1985 through 2009.

[No., number; USGS, U.S. Geological Survey; PP, physical properties; N, nutrients; CHL, chlorophyll; MI, major ions; TE, trace elements; OC, organic carbon; SI, stable isotopes; CDPHE, Colorado Department of Public Health and Environment; CB, coliform bacteria. Subwatershed definitions: Yampa River subwatershed 1, Yampa River and tributaries upstream from Chuck Lewis State Wildlife Area; Yampa River subwatershed 2, Yampa River and tributaries from Chuck Lewis State Wildlife Area to Elk River confluence; Elk River subwatershed, Elk River and tributaries; Elkhead Creek subwatershed, Elkhead Creek and tributaries. The Upper Yampa River water-quality database is available at *http //rmgsc.cr.usgs.gov/cwqdr/Yampa/index.shtml*]

Site name in Upper Yampa River watershed water-quality database	Source of data	Site identifier	Latitude	Longitude	Subwatershed
ELKHEAD RESERVOIR SITE 1A	USGS	403507107214900	40.585	−107.364	Elkhead Creek subwatershed
ELKHEAD RESERVOIR SITE 1B	USGS	403506107214500	40.585	−107.363	Elkhead Creek subwatershed
ELKHEAD RESERVOIR SITE 2A	USGS	403439107223800	40.577	−107.378	Elkhead Creek subwatershed
ELKHEAD RESERVOIR SITE 2B	USGS	403437107223300	40.577	−107.376	Elkhead Creek subwatershed
ELKHEAD RESERVOIR SITE 2C	USGS	403435107222900	40.576	−107.375	Elkhead Creek subwatershed
ELKHEAD RESERVOIR SITE 3A	USGS	403336107230700	40.560	−107.386	Elkhead Creek subwatershed
ELKHEAD RESERVOIR SITE 3B	USGS	403333107230100	40.559	−107.384	Elkhead Creek subwatershed
ELKHEAD RESERVOIR SITE 3C	USGS	403331107225500	40.559	−107.383	Elkhead Creek subwatershed
LAKE ELBERT	USGS	403803106422500	40.634	−106.708	Yampa River subwatershed 2
LONG LAKE RESERVOIR	USGS	402833106412400	40.476	−106.691	Yampa River subwatershed 2
STAGECOACH RES, NR PHIPPSBURGH, 250M W OF DAM	CDPHE	STAGE01	40.285	−106.835	Yampa River subwatershed 1
STAGECOACH RESERVOIR, NR PHIPPSBURGH,INLET SIDE	CDPHE	STAGE02	40.276	−106.860	Yampa River subwatershed 1
STAGECOACH RESERVOIR AT DAM, COLORADO	USGS	401707106495800	40.285	−106.833	Yampa River subwatershed 1
STAGECOACH RESERVOIR NEAR INLET, NR OAK CREEK, CO.	USGS	401628106515500	40.274	−106.866	Yampa River subwatershed 1
STAGECOACH RESERVOIR NR DAM [UPPER and LOWER]	CDPHE	12812A, 12812B	40.285	−106.835	Yampa River subwatershed 1
STEAMBOAT LAKE NR DAM [UPPER and LOWER]	CDPHE	12866A, 12866B	40.793	−106.949	Elk River subwatershed

Appendix 4.   Description of selected lake and reservoir samlping sites in the Upper Yampa River watershed, Colorado, with type of water-quality data collected, period of water-quality record, and number of sample days, 1985 through 2009.—Continued

[No., number; USGS, U.S. Geological Survey; PP, physical properties; N, nutrients; CHL, chlorophyll; MI, major ions; TE, trace elements; OC, organic carbon; SI, stable isotopes; CDPHE, Colorado Department of Public Health and Environment; CB, coliform bacteria. Subwatershed definitions: Yampa River subwatershed 1, Yampa River and tributaries upstream from Chuck Lewis State Wildlife Area; Yampa River subwatershed 2, Yampa River and tributaries from Chuck Lewis State Wildlife Area to Elk River confluence; Elk River subwatershed, Elk River and tributaries; Elkhead Creek subwatershed, Elkhead Creek and tributaries. The Upper Yampa River water-quality database is available at *http //rmgsc.cr.usgs.gov/cwqdr/Yampa/index.shtml*]

Site name in Upper Yampa River watershed water-quality database	Type of water-quality data collected	Period of water-quality record	No. of sample days[1]
ELKHEAD RESERVOIR SITE 1A	PP	7/12/1995–8/2/2001	18
ELKHEAD RESERVOIR SITE 1B	PP,N,CHL	7/12/1995–8/2/2001	18
ELKHEAD RESERVOIR SITE 2A	PP	7/12/1995–8/2/2001	18
ELKHEAD RESERVOIR SITE 2B	PP,N,CHL	7/12/1995–8/2/2001	18
ELKHEAD RESERVOIR SITE 2C	PP	7/12/1995–8/2/2001	18
ELKHEAD RESERVOIR SITE 3A	PP	7/12/1995–8/2/2001	18
ELKHEAD RESERVOIR SITE 3B	PP,N,CHL	7/12/1995–8/2/2001	18
ELKHEAD RESERVOIR SITE 3C	PP	7/12/1995–8/2/2001	18
LAKE ELBERT	PP,MI,N,TE,OC,CHL,SI	7/18/1985–9/9/2009	74
LONG LAKE RESERVOIR	PP,MI,N,TE,OC,CHL,SI	7/16/1985–8/11/2005	64
STAGECOACH RES, NR PHIPPSBURGH, 250M W OF DAM	PP,MI,N,TE,CHL	10/10/1990	1
STAGECOACH RESERVOIR, NR PHIPPSBURGH,INLET SIDE	PP,MI,N,TE,CHL	10/10/1990	1
STAGECOACH RESERVOIR AT DAM, COLORADO	PP,TDS,MI,N,TE,OC,CB	4/26/1990–11/7/1992	20
STAGECOACH RESERVOIR NEAR INLET,NR OAK CREEK, CO.	PP,TDS,MI,N,TE,OC,CB	4/26/1990–11/7/1992	20
STAGECOACH RESERVOIR NR DAM [UPPER and LOWER]	PP,MI,N,TE,CHL	7/25/2006	1
STEAMBOAT LAKE NR DAM [UPPER and LOWER]	PP,MI,N,TE,CHL	7/25/2006	1

[1]Number of days in which water samples were collected at the site for a lake or reservoir. Samples collected on the same day but at different times or depths in the lake or reservoir are counted as one sample day.

Appendix 5.   Description of selected groundwater sampling sites in the Upper Yampa River watershed, Colorado, with geologic unit description, type of water-quality data collected, period of water-quality record, number of samples collected, and constituents with exceedances of Colorado Department of Public Health and Environment water-quality standards for groundwater, 1975 through 1989 and 1998

[No., number; CDPHE, Colorado Department of Public Health and Environment; USGS, U.S. Geological Survey; PP, physical properties; --, water-quality standard not exceeded; DS, dissolved solids; MI, major ions; N, nutrients; TE, trace elements; Fe, iron; Mn, manganese; NO_3+NO_2, nitrate plus nitrite; Zn, zinc; Sul, sulfate; Cd, cadmium; Ar, arsenic; R, radiochemical; Chl, chloride; Bo, boron; Pb, lead; SI, stable isotopes; OC, organic carbon; Fl, fluoride; Be, beryllium; Cu, copper; Se, selenium; Mo, molybdenum; CDOA, Colorado Department of Agriculture. The Upper Yampa River water-quality database is available at *http //rmgsc.cr.usgs.gov/cwqdr/Yampa/index.shtml*]

Site no. (see figure 15)	Site name in Upper Yampa River watershed water-quality database	Source of data	Site Identifier	Latitude	Longitude
180	SB00208529DAD1	USGS	400637106563601	40.110	−106.944
181	SB00208521CDD1	USGS	400716106555901	40.121	−106.934
182	SB00208518DDC1	USGS	400809106574301	40.136	−106.963
183	SB00208514CCB1	USGS	400819106541001	40.139	−106.903
184	SB00208517CBA	USGS	400845106572001	40.146	−106.956
185	SB00208511CAB1	USGS	400921106534601	40.156	−106.897
186	SB00208510ADC1	USGS	400930106542101	40.158	−106.906
187	SB00208306DBC1	USGS	401007106442101	40.169	−106.740
188	SB00308635CCD1	USGS	401053107005201	40.181	−107.015
189	SB00308528ACC1	USGS	401210106555401	40.203	−106.932
190	SB00308528ACC2	USGS	401210106555601	40.203	−106.933
191	SB00308516CCD1	USGS	401330106562001	40.225	−106.939
192	SB00308016CDB1	USGS	401340107030201	40.228	−107.051
193	SB00308517DBB1	USGS	401349106565501	40.230	−106.949
194	SB00308517ADC1	USGS	401352106564201	40.231	−106.944
195	SB00308517ADD2	USGS	401355106563501	40.232	−106.944
196	SB00308517ADD1	USGS	401355106563601	40.232	−106.944
197	SB00308509BDC1	USGS	401418106560301	40.238	−106.935
198	SB00308618BBA1	USGS	401425107052301	40.240	−107.090
199	SB00308503ADB1	USGS	401551106541201	40.264	−106.904
200	SB00308605ACB1	USGS	401552107035301	40.264	−107.065
201	SB00408531DDC1	USGS	401611106575001	40.270	−106.964
202	SB00408530ACD1	USGS	401729106575701	40.291	−106.966
203	SB00408729ADA1	USGS	401735107104201	40.293	−107.179
204	SB00408622CCC1	USGS	401754107020501	40.298	−107.035
205	SB00408622CDB1	USGS	401801107015101	40.300	−107.031
206	SB00408724DBD	USGS	401804107062101	40.301	−107.106
207	SB00408724BCB	USGS	401826107070401	40.307	−107.118
208	SB00408619BBD	USGS	401837107054501	40.310	−107.096
209	SB00408620AAA1	USGS	401842107032201	40.312	−107.057
210	SB00408614DCC1	USGS	401847107003101	40.313	−107.009
211	SB00408614CDD FOIDEL MIDDLE CRK 8	USGS	401847107003301	40.313	−107.011
212	SB00408517CCB1	USGS	401857106573101	40.316	−106.959
213	SB00408713AAD US GOVT	USGS	401904107060800	40.318	−107.103
214	SB00408713DAA	USGS	401904107060801	40.318	−107.103
215	SB00408616CBA FOIDEL MIDDLE CRK 3	USGS	401912107031300	40.320	−107.054
216	SB00408616CBA	USGS	401912107031301	40.320	−107.054
217	SB00408618BDA	USGS	401922107050701	40.323	−107.086
218	SB00408416BBB1	USGS	401931106492301	40.325	−106.824
219	SB00408612CBD1	USGS	401935106595401	40.326	−106.999
220	SB00408710CCD1	USGS	401939107091201	40.327	−107.154
221	SB00408611DCC	USGS	401941107002001	40.328	−107.006
222	SB00408709CBC DALE WELDON	USGS	401950107103300	40.331	−107.176
223	SB00408407DBB1	USGS	401959106510101	40.333	−106.851
224	SB00408407BDB1	USGS	402007106513001	40.335	−106.859
225	SB00408610BAD	USGS	402015107014501	40.337	−107.030
226	SB00408610BAD FOIDEL-MIDDLE CREEK 4	USGS	402015107015500	40.337	−107.033
227	SB00408512AAD1	USGS	402023106515201	40.340	−106.865
228	SB00508634CAB1	USGS	402041107013301	40.345	−107.026
229	SB00508433DBB1	USGS	402049106485501	40.347	−106.816

Appendix 5. Description of selected groundwater sampling sites in the Upper Yampa River watershed, Colorado, with geologic unit description, type of water-quality data collected, period of water-quality record, number of samples collected, and constituents with exceedances of Colorado Department of Public Health and Environment water-quality standards for groundwater, 1975 through 1989 and 1998.—Continued

[No., number; CDPHE, Colorado Department of Public Health and Environment; USGS, U.S. Geological Survey; PP, physical properties; --, water-quality standard not exceeded; DS, dissolved solids; MI, major ions; N, nutrients; TE, trace elements; Fe, iron; Mn, manganese; NO_3+NO_2, nitrate plus nitrite; Zn, zinc; Sul, sulfate; Cd, cadmium; Ar, arsenic; R, radiochemical; Chl, chloride; Bo, boron; Pb, lead; SI, stable isotopes; OC, organic carbon; Fl, fluoride; Be, beryllium; Cu, copper; Se, selenium; Mo, molybdenum; CDOA, Colorado Department of Agriculture. The Upper Yampa River water-quality database is available at *http //rmgsc.cr.usgs.gov/cwqdr/Yampa/index.shtml*]

Site no. (see figure 15)	Geologic unit description	Type of water-quality data collected[1]	Period of water-quality record (calendar year)	No. of sample days	No. of samples[2]	Constituent with exceedance of CDPHE water-quality standard[3]
180	Alluvium, flood plain	PP	1978	1	1	--
181	Mancos Shale	PP, DS, MI, N, TE	1978	1	1	--
182	Curtis Formation of San Rafael Group	PP, DS, MI, N, TE	1978	1	1	--
183	Alluvium, flood plain	PP	1978	1	1	--
184	Unknown	PP, DS, MI, TE	1981	1	1	--
185	Mancos Shale	PP, DS, MI, N, TE	1978	1	1	Fe, Mn
186	Alluvium, flood plain	PP, DS, MI, N, TE	1978	1	1	--
187	Browns Park Formation	PP	1978	1	1	--
188	Mancos Shale	PP, DS, MI, N, TE	1978	1	1	--
189	Browns Park Formation	PP	1978	1	1	--
190	Browns Park Formation	PP	1978	1	1	--
191	Browns Park Formation	PP, DS, MI, N, TE	1978	1	1	--
192	Alluvium, flood plain	PP, DS, MI, N, TE	1975	1	1	--
193	Browns Park Formation	PP	1978	1	1	--
194	Browns Park Formation	PP	1978	1	1	--
195	Browns Park Formation	PP	1978	1	1	--
196	Alluvium, flood plain	PP	1978	1	1	--
197	Mancos Shale	PP, DS, MI, N, TE	1978	1	1	NO_3+NO_2, Zn
198	Alluvium, flood plain	PP	1978	1	1	pH
199	Unknown	PP	1975	1	1	--
200	Alluvium, flood plain	PP	1975	1	1	--
201	Unknown	PP, DS, MI, N, TE	1975, 1978	2	2	--
202	Unknown	PP, DS, MI, N, TE	1975	2	2	Sul, Fe, Mn
203	Unknown	PP	1975	1	1	--
204	Alluvium, flood plain	PP	1975	1	1	--
205	Mesaverde Group	PP, DS, MI, N, TE	1978	1	1	--
206	Mesaverde Group	PP, DS, MI, N, TE	1977, 1980	2	2	Sul, Mn (2)
207	Mesaverde Group	PP, DS, MI, N, TE	1977, 1980	2	2	Fe, Mn
208	Mesaverde Group	PP, DS, MI, N, TE	1977, 1980	3	3	Mn (2)
209	Unknown	PP	1975	1	1	--
210	Mesaverde Group	PP, DS, MI, N, TE	1977	1	1	Mn
211	Unknown	PP, DS, MI, N, TE	1975	1	1	pH
212	Alluvium, flood plain	PP	1975	1	1	--
213	Unknown	PP, DS, MI, N, TE	1975–76	2	2	Sul (2), Fe (2), Mn
214	Mesaverde Group	PP, DS, MI, N, TE	1977, 1980	2	2	Sul (2), Fe (2), Cd, Mn
215	Unknown	PP, DS, MI, N, TE	1975	1	1	pH
216	Mesaverde Group	PP, DS, MI, N, TE	1977	1	1	--
217	Mesaverde Group	PP, DS, MI, N, TE	1977	1	1	pH, Fe, Mn
218	Browns Park Formation	PP	1978	1	1	--
219	Mesaverde Group	PP, DS, MI, N, TE	1980	1	1	pH
220	Alluvium, flood plain	PP	1975	1	1	--
221	Unknown	PP, DS, MI, N, TE	1980	1	1	Mn
222	Mancos Shale	PP, DS, MI, N, TE	1975	1	1	Sul
223	Browns Park Formation	PP, DS, MI, N, TE	1978	1	1	--
224	Browns Park Formation	PP	1978	1	1	--
225	Mesaverde Group	PP, DS, MI, N, TE	1977	1	1	Mn
226	Valley-fill deposits	PP, DS, MI, N, TE	1975	1	1	Fe, Mn
227	Browns Park Formation	PP	1978	1	1	--
228	Unknown	PP, DS, MI, N, TE	1975	2	2	Mn
229	Alluvium, flood plain	PP	1978	1	1	--

Appendix 5.   Description of selected groundwater sampling sites in the Upper Yampa River watershed, Colorado, with geologic unit description, type of water-quality data collected, period of water-quality record, number of samples collected, and constituents with exceedances of Colorado Department of Public Health and Environment water-quality standards for groundwater, 1975 through 1989 and 1998.—Continued

[No., number; CDPHE, Colorado Department of Public Health and Environment; USGS, U.S. Geological Survey; PP, physical properties; --, water-quality standard not exceeded; DS, dissolved solids; MI, major ions; N, nutrients; TE, trace elements; Fe, iron; Mn, manganese; NO_3+NO_2, nitrate plus nitrite; Zn, zinc; Sul, sulfate; Cd, cadmium; Ar, arsenic; R, radiochemical; Chl, chloride; Bo, boron; Pb, lead; SI, stable isotopes; OC, organic carbon; Fl, fluoride; Be, beryllium; Cu, copper; Se, selenium; Mo, molybdenum; CDOA, Colorado Department of Agriculture. The Upper Yampa River water-quality database is available at *http //rmgsc.cr.usgs.gov/cwqdr/Yampa/index.shtml*]

Site no. (see figure 15)	Site name in Upper Yampa River watershed water-quality database	Source of data	Site Identifier	Latitude	Longitude
230	SB00508433DBB GLEN BARBER	USGS	402050106475500	40.347	−106.799
231	SB00508636BDD1	USGS	402056106590901	40.349	−106.986
232	SB00508622DCC1 USGS 402106107011801	USGS	402106107011801	40.352	−107.022
233	SB00508630CCC1	USGS	402112107050801	40.353	−107.086
234	SB00508636BAA2	USGS	402113106590501	40.354	−106.985
235	SB00508629CDD1	USGS	402114107034300	40.354	−107.063
236	SB00508629CDD2	USGS	402114107034301	40.354	−107.063
237	SB00508629DCC1	USGS	402118107033101	40.355	−107.059
238	SB00508526CCC1	USGS	402120106535201	40.356	−106.898
239	SB00408525DDD1	USGS	402121106515301	40.356	−106.865
240	SB00508629DDB	USGS	402124107031801	40.357	−107.056
241	SB00508625DDB1	USGS	402128106583901	40.358	−106.978
242	SB00508525DAD1	USGS	402132106515001	40.359	−106.864
243	SB00508527ACC1	USGS	402146106543401	40.363	−106.910
244	SB00508627BCA1	USGS	402149107013801	40.364	−107.028
245	SB00508627BBC1	USGS	402156107015201	40.366	−107.032
246	SB00508527ABC1	USGS	402157106543101	40.366	−106.909
247	SB00508025AAC1	USGS	402157106583801	40.366	−106.978
248	SB00408529BAC1	USGS	402158106570701	40.366	−106.953
249	SB00508628ABB1	USGS	402202107022101	40.367	−107.040
250	SB00508809CDB1	USGS	402202107160801	40.367	−107.270
251	SB00508627BAA	USGS	402204107012201	40.368	−107.023
252	SB00508628BAA1 USGS 402204107022801	USGS	402204107022801	40.368	−107.042
253	SB00508627BAB1	USGS	402205107013201	40.368	−107.026
254	SB00508621CDD	USGS	402209107023101	40.369	−107.043
255	SB00508622DCC1 USGS 402212107011801	USGS	402212107011801	40.370	−107.022
256	SB005087019CDA2	USGS	402217107113601	40.371	−107.194
257	SB00508719CDA1	USGS	402222107113001	40.373	−107.192
258	SB00508719ABA2	USGS	402231107111601	40.382	−107.188
259	SB00508719ABA1	USGS	402231107111602	40.382	−107.188
260	SB00508720BBA2	USGS	402231107111612	40.382	−107.179
261	SB00508717BCC1	USGS	402231107111616	40.392	−107.179
262	SB00508717ADC1	USGS	402231107111617	40.392	−107.166
263	SB00508719ABB1	USGS	402231107111630	40.382	−107.189
264	SB00508719ABB2	USGS	402231107111631	40.382	−107.189
265	SB00508719ABB3	USGS	402231107111632	40.382	−107.189
266	SB00508719ABB4	USGS	402231107111633	40.382	−107.189
267	SB00508719DBA3	USGS	402231107111634	40.375	−107.188
268	SB00508719DBA5	USGS	402231107111636	40.375	−107.188
269	SB00508719DBA6	USGS	402231107111637	40.375	−107.188
270	SB00508718CAC1	USGS	402231107111646	40.388	−107.196
271	SB00508719CCB3	USGS	402231107111653	40.376	−107.194
272	SB00508719CDC2	USGS	402231107111657	40.369	−107.195
273	SB00508621BCC	USGS	402236107025301	40.377	−107.049
274	SB00508523ADC1	USGS	402237106530701	40.377	−106.886
275	SB00508620BCB	USGS	402239107040301	40.377	−107.068
276	SB00508820ACA1	USGS	402244107165001	40.379	−107.281
277	SB00508822BBC1 USGS 402250107151801	USGS	402250107151801	40.381	−107.256
278	SB00508822BBC1 USGS 402250107151802	USGS	402250107151802	40.381	−107.256
279	SB00508621AAA	USGS	402257107015301	40.382	−107.032
280	SB00508420BAB1	USGS	402259106501501	40.383	−106.838

Appendix 5. Description of selected groundwater sampling sites in the Upper Yampa River watershed, Colorado, with geologic unit description, type of water-quality data collected, period of water-quality record, number of samples collected, and constituents with exceedances of Colorado Department of Public Health and Environment water-quality standards for groundwater, 1975 through 1989 and 1998.—Continued

[No., number; CDPHE, Colorado Department of Public Health and Environment; USGS, U.S. Geological Survey; PP, physical properties; --, water-quality standard not exceeded; DS, dissolved solids; MI, major ions; N, nutrients; TE, trace elements; Fe, iron; Mn, manganese; NO_3+NO_2, nitrate plus nitrite; Zn, zinc; Sul, sulfate; Cd, cadmium; Ar, arsenic; R, radiochemical; Chl, chloride; Bo, boron; Pb, lead; SI, stable isotopes; OC, organic carbon; Fl, fluoride; Be, beryllium; Cu, copper; Se, selenium; Mo, molybdenum; CDOA, Colorado Department of Agriculture. The Upper Yampa River water-quality database is available at *http //rmgsc.cr.usgs.gov/cwqdr/Yampa/index.shtml*]

Site no. (see figure 15)	Geologic unit description	Type of water-quality data collected[1]	Period of water-quality record (calendar year)	No. of sample days	No. of samples[2]	Constituent with exceedance of CDPHE water-quality standard[3]
230	Unknown	PP, DS, MI, N, TE	1975	1	1	--
231	Mesaverde Group	PP, DS, MI, N, TE	1975, 1978	3	3	--
232	Lewis Shale	PP	1975	1	1	--
233	Unknown	PP, DS, MI, N, TE	1975	2	2	--
234	Mesaverde Group	PP, DS, MI, N, TE	1980	1	1	--
235	Mesaverde Group	PP, DS, MI, N, TE	1975	1	1	--
236	Mesaverde Group	PP, DS, MI, N, TE	1975, 1977	2	2	pH, Cd
237	Mesaverde Group	PP, DS, MI, N, TE	1975	2	2	--
238	Alluvium, flood plain	PP, DS, MI, N, TE	1975	2	2	Ar
239	Browns Park Formation	PP	1978	1	1	--
240	Mesaverde Group	PP, DS, MI, N, TE	1975	2	2	Sul
241	Unknown	PP	1975	1	1	--
242	Browns Park Formation	PP	1978	1	1	--
243	Alluvium, flood plain	PP	1978	1	1	--
244	Lewis Shale	PP	1975	1	1	--
245	Lewis Shale	PP, DS, MI, N, TE	1975	2	2	Fe, Mn
246	Alluvium, flood plain	PP	1975	1	1	--
247	Unknown	PP	1975	1	1	--
248	Mancos Shale	PP	1978	1	1	--
249	Alluvium, flood plain	PP, DS, MI, N, TE	1975	2	2	Mn
250	Alluvium, flood plain	PP	1975	1	1	--
251	Alluvium, flood plain	PP, DS, MI, N, TE	1976, 1980	3	3	NO_3+NO_2, Fe, Mn (2)
252	Mesaverde Group	PP, DS, MI, N, TE	1975	1	1	--
253	Lewis Shale	PP, DS, MI, N, TE	1975	2	2	Mn
254	Mesaverde Group	PP, DS, MI, N, TE	1976–77, 1980	3	3	pH (2), Cd
255	Valley-fill deposits	PP, DS, MI, N, TE	1975	1	1	--
256	Mesaverde Group	PP	1978	1	1	--
257	Unknown	PP, DS, MI, N, TE	1975	2	2	Sul, Fe, Mn
258	Mesaverde Group	PP, DS, MI, N, TE, R	1980–81	4	4	Sul (4), Fe (2), Mn (3)
259	Mesaverde Group	PP, DS, MI, N, TE, R	1980–83	9	9	Sul (2)
260	Mesaverde Group	PP, DS, MI, N, TE, R	1980–83	9	9	pH (2), Mn (6)
261	Unknown	PP, DS, MI, N, TE, R	1980–82	7	7	Mn (6)
262	Unknown	PP, DS, MI, N, TE, R	1980–82	5	5	Fe (2), Mn (5)
263	Unknown	PP, DS, MI, N, TE	1981	1	1	Sul, Mn
264	Unknown	PP, DS, MI, N, TE	1981	1	1	Sul, Mn
265	Unknown	PP, DS, MI, N, TE	1981	1	1	Sul, Mn
266	Unknown	PP, DS, MI, N, TE	1981	1	1	Sul, Mn
267	Unknown	PP, DS, MI, N, TE, R	1980–83	9	9	Sul (4), Fe, Mn (8)
268	Unknown	PP, DS, MI, N, TE, R	1980–82	6	6	pH (5), Chl (6), Fe, Mn
269	Unknown	PP, DS, MI, N, TE, R	1982–83	7	7	pH (4), Sul (5), Chl (2)
270	Unknown	PP, DS, MI, N, TE, R	1980–83	12	12	Sul (4), Mn (2)
271	Unknown	PP, DS, MI, N, TE, R	1981–82	6	6	Sul (6), Fe, Mn (6)
272	Unknown	PP, DS, MI, N, TE, R	1980–82	6	6	Fe (2), Mn (4)
273	Mesaverde Group	PP, DS, MI, N, TE	1976–77, 1980	3	3	Sul, Chl (2), Bo
274	Browns Park Formation	PP	1978	1	1	--
275	Mesaverde Group	PP, DS, MI, N, TE	1976–77	2	2	--
276	Unknown	PP, DS, MI, N, TE, R	1980–81	4	4	Fe, Mn (3)
277	Mesaverde Group	PP, DS, MI, N, TE	1975	1	1	Fe, Zn
278	Unknown	PP	1975	1	1	--
279	Mesaverde Group	PP, DS, MI, N, TE	1976–77, 1980	3	3	pH (2)
280	Browns Park Formation	PP	1978	1	1	--

Appendix 5.   Description of selected groundwater sampling sites in the Upper Yampa River watershed, Colorado, with geologic unit description, type of water-quality data collected, period of water-quality record, number of samples collected, and constituents with exceedances of Colorado Department of Public Health and Environment water-quality standards for groundwater, 1975 through 1989 and 1998.—Continued

[No., number; CDPHE, Colorado Department of Public Health and Environment; USGS, U.S. Geological Survey; PP, physical properties; --, water-quality standard not exceeded; DS, dissolved solids; MI, major ions; N, nutrients; TE, trace elements; Fe, iron; Mn, manganese; NO_3+NO_2, nitrate plus nitrite; Zn, zinc; Sul, sulfate; Cd, cadmium; Ar, arsenic; R, radiochemical; Chl, chloride; Bo, boron; Pb, lead; SI, stable isotopes; OC, organic carbon; Fl, fluoride; Be, beryllium; Cu, copper; Se, selenium; Mo, molybdenum; CDOA, Colorado Department of Agriculture. The Upper Yampa River water-quality database is available at *http //rmgsc.cr.usgs.gov/cwqdr/Yampa/index.shtml*]

Site no. (see figure 15)	Site name in Upper Yampa River watershed water-quality database	Source of data	Site Identifier	Latitude	Longitude
281	SB00508915CAB	USGS	402303107215701	40.384	−107.366
282	SB00508913ACC	USGS	402316107182301	40.388	−107.307
283	SB00508416DBB1	USGS	402324106485701	40.390	−106.816
284	SB00508515CBA1	USGS	402324106545501	40.390	−106.916
285	SB00508613ACC FOIDEL-MIDDLE CRK 5	USGS	402327106590000	40.391	−106.984
286	SB00508613ACC	USGS	402327106590001	40.391	−106.984
287	SB00508816CAB1	USGS	402327107161301	40.391	−107.271
288	SB00508515ADC1	USGS	402329106541001	40.391	−106.903
289	SB00508518ACA1	USGS	402334106574601	40.393	−106.963
290	SB00508518ACA	USGS	402336106574400	40.393	−106.963
291	SB00508417BCB RICHARD RUBISH	USGS	402338106503600	40.394	−106.844
292	SB00508912CCD1	USGS	402344107195001	40.396	−107.331
293	SB00508613ABB FOIDEL MIDDLE CRK 6	USGS	402346106590000	40.396	−106.984
294	SB00508613ABB	USGS	402346106590001	40.396	−106.984
295	SB00508514BBB1	USGS	402349106535201	40.397	−106.898
296	SB00508516AAA1	USGS	402351106482201	40.397	−106.807
297	SB00508808CDC	USGS	402356107171701	40.399	−107.289
298	SB00508511CCC1	USGS	402357106535801	40.399	−106.900
299	SB00508511CCC BETTY PUGH	USGS	402358106535900	40.399	−106.900
300	SCL287	USGS	402359107054401	40.400	−107.096
301	SCI287	USGS	402359107054402	40.400	−107.096
302	SCW287	USGS	402359107054403	40.400	−107.096
303	SCU287	USGS	402359107054404	40.400	−107.096
304	SCS1487-62	USGS	402401107054701	40.400	−107.097
305	SCS2487-63	USGS	402401107054702	40.400	−107.097
306	SCS3487-64	USGS	402401107054703	40.400	−107.097
307	SB00508612DBB1 USGS 402414106585701	USGS	402414106585701	40.404	−106.983
308	SB00508807CBB1	USGS	402415107184401	40.404	−107.313
309	SB00508509ADC GARY BRENNEMAN	USGS	402423106552500	40.406	−106.924
310	SB00508407ABD1	USGS	402436106510101	40.410	−106.851
311	SBL287	USGS	402437107051701	40.410	−107.089
312	SBI287	USGS	402437107051702	40.410	−107.089
313	SBW287	USGS	402437107051703	40.410	−107.089
314	SBU287	USGS	402437107051704	40.410	−107.089
315	SB00500508BBC1	USGS	402438106572001	40.411	−106.956
316	SB00508612BAC	USGS	402440106591001	40.411	−106.987
317	SSL287	USGS	402454107071601	40.415	−107.122
318	SSI287	USGS	402454107071602	40.415	−107.122
319	SSW287	USGS	402454107071603	40.415	−107.122
320	SSU487	USGS	402454107071604	40.415	−107.122
321	SSU187	USGS	402454107071605	40.415	−107.122
322	SSD487	USGS	402454107071606	40.415	−107.122
323	SZL287	USGS	402455107073001	40.415	−107.126
324	SZI287	USGS	402455107073002	40.415	−107.126
325	SZW487	USGS	402455107073003	40.415	−107.126
326	SZU287	USGS	402455107073004	40.415	−107.126
327	SSS1487-59	USGS	402456107071101	40.416	−107.120

Appendix 5.   Description of selected groundwater sampling sites in the Upper Yampa River watershed, Colorado, with geologic unit description, type of water-quality data collected, period of water-quality record, number of samples collected, and constituents with exceedances of Colorado Department of Public Health and Environment water-quality standards for groundwater, 1975 through 1989 and 1998.—Continued

[No., number; CDPHE, Colorado Department of Public Health and Environment; USGS, U.S. Geological Survey; PP, physical properties; --, water-quality standard not exceeded; DS, dissolved solids; MI, major ions; N, nutrients; TE, trace elements; Fe, iron; Mn, manganese; NO_3+NO_2, nitrate plus nitrite; Zn, zinc; Sul, sulfate; Cd, cadmium; Ar, arsenic; R, radiochemical; Chl, chloride; Bo, boron; Pb, lead; SI, stable isotopes; OC, organic carbon; Fl, fluoride; Be, beryllium; Cu, copper; Se, selenium; Mo, molybdenum; CDOA, Colorado Department of Agriculture. The Upper Yampa River water-quality database is available at *http //rmgsc.cr.usgs.gov/cwqdr/Yampa/index.shtml*]

Site no. (see figure 15)	Geologic unit description	Type of water-quality data collected[1]	Period of water-quality record (calendar year)	No. of sample days	No. of samples[2]	Constituent with exceedance of CDPHE water-quality standard[3]
281	Unknown	PP, DS, MI, N, TE	1980	1	1	Sul, Fe
282	Unknown	PP, DS, MI, N, TE	1980	1	1	--
283	Alluvium, flood plain	PP	1978	1	1	--
284	Browns Park Formation	PP	1978	1	1	--
285	Unknown	PP, DS, MI, N, TE	1975	1	1	Fe
286	Mesaverde Group	PP, DS, MI, N, TE	1977	1	1	Chl, Pb
287	Unknown	PP, DS, MI, N, TE, R	1980–81	4	4	Sul (4), Fe (3), Mn (4)
288	Browns Park Formation	PP	1978	1	1	--
289	Mesaverde Group	PP	1978	1	1	--
290	Unknown	PP, DS, MI, N, TE	1975	1	1	--
291	Unknown	PP, DS, MI, N, TE	1975	1	1	Chl
292	Unknown	PP	1975	1	1	--
293	Unknown	PP, DS, MI, N, TE	1975	1	1	Sul, Mn
294	Mesaverde Group	PP, DS, MI, N, TE	1977	1	1	pH, Sul
295	Browns Park Formation	PP	1978	1	1	--
296	Alluvium, flood plain	PP	1978	1	1	--
297	Unknown	PP, DS, MI, N, TE	1976–77	2	2	pH, Sul (2), Fe
298	Browns Park Formation	PP	1978	1	1	--
299	Unknown	PP, DS, MI, N, TE	1975	1	1	--
300	Unknown	PP, DS, MI, N, TE	1987–89	9	9	Sul (9), Fe (3), Mn (9)
301	Unknown	PP, DS, MI, N, TE	1987–89	9	9	Sul (7), Mn (2)
302	Unknown	PP, DS, MI, N, TE, R, SI	1987–89	9	9	pH (2), Sul (8), Mn (2)
303	Unknown	PP, DS, MI, N, TE, OC, R, SI	1987–89	10	11	pH (5), Sul
304	Unknown	PP, DS, MI, N, TE, OC, R, SI	1988–89	9	11	Sul (9), NO_3+NO_2, Cd, Mn (9)
305	Unknown	PP, DS, MI, N, TE	1988	4	4	Sul (4), Cd, Mn (4)
306	Unknown	PP, DS, MI, N, TE	1987–88	5	5	Sul (5), Mn (5)
307	Mesaverde Group	PP, DS, MI, N, TE	1975	1	1	Fe
308	Lewis Shale	PP	1978	1	1	--
309	Unknown	PP, DS, MI, N, TE	1975	1	1	--
310	Browns Park Formation	PP	1978	1	1	--
311	Unknown	PP, DS, MI, N, TE	1987–89	12	12	Cd, Fe, Mn (8)
312	Unknown	PP, DS, MI, N, TE	1987–89	13	15	Sul (3), NO_3+NO_2, Cd, Mn (13)
313	Unknown	PP, DS, MI, N, TE, R, SI	1987–89	14	15	Mn (10)
314	Unknown	PP, DS, MI, N, TE, R, SI	1987–89	12	12	Sul, Mn (8)
315	Browns Park Formation	PP	1975	1	1	--
316	Unknown	PP, DS, MI, N, TE	1980	1	1	--
317	Unknown	PP, DS, MI, N, TE	1987–89	10	12	Sul (10), Fe (2), Mn (10)
318	Unknown	PP, DS, MI, N, TE	1987–89	9	9	Sul (9), Cd
319	Unknown	PP, DS, MI, N, TE, R, SI	1987–89	9	9	pH, Sul (9), Fe (2), Mn (9)
320	Unknown	PP, DS, MI, N, TE, R, SI	1987–89	10	12	Sul (10), Cd, Mn (2)
321	Unknown	PP, DS, MI, N, TE, R, SI	1987–89	9	9	pH (8), Sul (5)
322	Unknown	PP, DS, MI, N, TE, OC, R, SI	1987–89	9	9	pH (9)
323	Unknown	PP, DS, MI, N, TE	1987–89	9	9	pH, Cd
324	Unknown	PP, DS, MI, N, TE	1987–89	10	12	pH (7)
325	Unknown	PP, DS, MI, N, TE, R, SI	1987–89	6	6	pH (2), Sul (2), Fe, Mn (2)
326	Unknown	PP, DS, MI, N, TE, R, SI	1987–89	9	9	pH (3), Fl (2)
327	Unknown	PP, DS, MI, N, TE	1987–88	4	4	Sul (4), NO_3+NO_2, Cd, Mn (4)

Appendix 5.   Description of selected groundwater sampling sites in the Upper Yampa River watershed, Colorado, with geologic unit description, type of water-quality data collected, period of water-quality record, number of samples collected, and constituents with exceedances of Colorado Department of Public Health and Environment water-quality standards for groundwater, 1975 through 1989 and 1998.—Continued

[No., number; CDPHE, Colorado Department of Public Health and Environment; USGS, U.S. Geological Survey; PP, physical properties; --, water-quality standard not exceeded; DS, dissolved solids; MI, major ions; N, nutrients; TE, trace elements; Fe, iron; Mn, manganese; NO_3+NO_2, nitrate plus nitrite; Zn, zinc; Sul, sulfate; Cd, cadmium; Ar, arsenic; R, radiochemical; Chl, chloride; Bo, boron; Pb, lead; SI, stable isotopes; OC, organic carbon; Fl, fluoride; Be, beryllium; Cu, copper; Se, selenium; Mo, molybdenum; CDOA, Colorado Department of Agriculture. The Upper Yampa River water-quality database is available at *http //rmgsc.cr.usgs.gov/cwqdr/Yampa/index.shtml*]

Site no. (see figure 15)	Site name in Upper Yampa River watershed water-quality database	Source of data	Site Identifier	Latitude	Longitude
328	SSS2487-60	USGS	402456107071102	40.416	−107.120
329	SSS3487-61	USGS	402456107071103	40.416	−107.120
330	SB00508804DBC1	USGS	402459107154901	40.416	−107.264
331	SB00508505ADD1	USGS	402514106561501	40.421	−106.938
332	SB00508603ADC1	USGS	402515107005401	40.421	−107.016
333	SB00508601BDA1	USGS	402517106590701	40.421	−106.986
334	SB00508904BBA1	USGS	402526107231601	40.424	−107.388
335	SB00608531CCC1 USGS 402535106582301	USGS	402535106582301	40.426	−106.974
336	SB00608531CCC1 USGS 402535106582501	USGS	402535106582501	40.426	−106.974
337	SB00608932DDB	USGS	402535107234401	40.426	−107.396
338	SB00508503BBB1	USGS	402545106481901	40.429	−106.806
339	SB6N87W34DDB1	USGS	402545107073901	40.429	−107.128
340	SB00608431DDB1	USGS	402551106505201	40.431	−106.848
341	SB00608833DBB	USGS	402600107160001	40.433	−107.267
342	SB00608836BDC1	USGS	402609107124801	40.436	−107.214
343	SB00608734ADA1	USGS	402620107072928	40.437	−107.127
344	SB00608734ADA2	USGS	402620107072929	40.437	−107.128
345	SB00608734ADB3	USGS	402620107072930	40.437	−107.128
346	SB00608734ACA1	USGS	402620107072932	40.438	−107.131
347	SB00608734ACA2	USGS	402620107072933	40.437	−107.130
348	SB00508703DAC1	USGS	402620107072935	40.418	−107.128
349	SB00608722DDD1 GW-S21,NW OF FLY ASHPIT	USGS	402620107072939	40.456	−107.128
350	SB00608734DDB1	USGS	402620107072950	40.430	−107.129
351	SB00608734DDB2	USGS	402620107072951	40.430	−107.125
352	SB00508703DAA1	USGS	402620107072952	40.420	−107.119
353	SB00508711BDB1	USGS	402620107072953	40.408	−107.119
354	SB00508711BDB2	USGS	402620107072955	40.408	−107.119
355	SB00608734BAB1 AW-1SENECA	USGS	402620107072956	40.441	−107.138
356	SB00608731BBB JIM ROWLEY	USGS	402627107115900	40.441	−107.200
357	SB00608433BAD CYRIL BARBER	USGS	402631106485700	40.442	−106.816
358	SB00608727DCA	USGS	402633107080101	40.442	−107.134
359	SB00608429DDD LOY ARDREY	USGS	402639106493600	40.444	−106.827
360	SB00608728DCA1	USGS	402641107090201	40.445	−107.151
361	SB00608828DAD1	USGS	402643107153301	40.445	−107.260
362	SB00608625DBC1	USGS	402646106585501	40.446	−106.983
363	SB00608625DBD ARNOLD LIESKE	USGS	402647106585000	40.446	−106.981
364	SB00608625BDB1	USGS	402709106591201	40.452	−106.987
365	SB00608930BAB1	USGS	402709107252501	40.452	−107.424
366	SB00608625BAC1	USGS	402710106591401	40.453	−106.988
367	SB00608720AAC1	USGS	402804107095201	40.468	−107.165
368	SB00608623ABA1	USGS	402809106595501	40.469	−106.999
369	SB00608509DDD1 USGS 402810106551001	USGS	402810106551001	40.469	−106.920
370	SB00608623BBA1	USGS	402815107003001	40.471	−107.009
371	SB00608813ADD2	USGS	402842107120001	40.478	−107.201
372	SB00608815BDC1	USGS	402843107150601	40.479	−107.252
373	SB00608815BCD1	USGS	402843107151201	40.479	−107.254
374	SB00608813ADD1	USGS	402845107115901	40.479	−107.200
375	SB00608814ADA1	USGS	402852107125801	40.481	−107.217
376	SB00608813ABC1	USGS	402857107122601	40.482	−107.208

Appendix 5.   Description of selected groundwater sampling sites in the Upper Yampa River watershed, Colorado, with geologic unit description, type of water-quality data collected, period of water-quality record, number of samples collected, and constituents with exceedances of Colorado Department of Public Health and Environment water-quality standards for groundwater, 1975 through 1989 and 1998.—Continued

[No., number; CDPHE, Colorado Department of Public Health and Environment; USGS, U.S. Geological Survey; PP, physical properties; --, water-quality standard not exceeded; DS, dissolved solids; MI, major ions; N, nutrients; TE, trace elements; Fe, iron; Mn, manganese; NO_3+NO_2, nitrate plus nitrite; Zn, zinc; Sul, sulfate; Cd, cadmium; Ar, arsenic; R, radiochemical; Chl, chloride; Bo, boron; Pb, lead; SI, stable isotopes; OC, organic carbon; Fl, fluoride; Be, beryllium; Cu, copper; Se, selenium; Mo, molybdenum; CDOA, Colorado Department of Agriculture. The Upper Yampa River water-quality database is available at *http //rmgsc.cr.usgs.gov/cwqdr/Yampa/index.shtml*]

Site no. (see figure 15)	Geologic unit description	Type of water-quality data collected[1]	Period of water-quality record (calendar year)	No. of sample days	No. of samples[2]	Constituent with exceedance of CDPHE water-quality standard[3]
328	Unknown	PP, DS, MI, N, TE	1987–88	5	5	Sul (5), NO_3+NO_2 (3), Cd, Mn (5)
329	Unknown	PP, DS, MI, N, TE, OC, R, SI	1987–89	10	12	Sul (10), NO_3+NO_2 (5), Be (2), Cd, Mn (10)
330	Unknown	PP, DS, MI, N, TE	1975	2	2	Sul, Mn
331	Browns Park Formation	PP, DS, MI, N, TE	1978	1	1	NO_3+NO_2
332	Mesaverde Group	PP, MI, N, TE	1980	1	1	pH
333	Alluvium, flood plain	PP	1978	1	1	pH
334	Unknown	PP, DS, MI, N, TE	1975	2	2	Sul, Fe, Mn
335	Alluvium, flood plain	PP	1978	1	1	pH
336	Alluvium, flood plain	PP	1978	1	1	pH
337	Unknown	PP	1975	1	1	--
338	Browns Park Formation	PP	1978	1	1	pH
339	Mesaverde Group	PP, DS, MI, N, TE	1977	1	1	pH, Sul
340	Browns Park Formation	PP	1977	1	1	--
341	Unknown	PP, DS, MI, N, TE	1976–77	3	3	pH (2), Fe
342	Mesaverde Group	PP, DS, MI, N, TE	1978	1	1	NO_3+NO_2
343	Mesaverde Group	PP, DS, MI, N, TE, R	1980–81, 1983	6	6	Sul (6), Bo (2), Fe, Mn (5)
344	Mesaverde Group	PP, DS, MI, N, TE, R	[4]1980–86	12	12	pH, Sul (12), Mn (11)
345	Mesaverde Group	PP, DS, MI, N, TE, R	[4]1980–86	13	13	pH, Sul (13), Mn (12)
346	Mesaverde Group	PP, DS, MI, TE	1984	1	1	Sul
347	Mesaverde Group	PP, DS, MI, N, TE, R	[4]1979–86	14	14	pH (3), Sul (14), Mn (5)
348	Mesaverde Group	PP, DS, MI, N, TE	1979–80	2	2	pH (2)
349	Unknown	PP, DS, MI, N, TE	1980	1	1	Sul, Mn
350	Mesaverde Group	PP, DS, MI, N, TE, R	1979–82	6	6	Sul
351	Mesaverde Group	PP, DS, MI, N, TE, R	1979–82	5	5	pH, Sul (4), Fe (2), Mn (5)
352	Mesaverde Group	PP, DS, MI, N, TE, OC, R	1979–83	10	10	pH
353	Unknown	PP, DS, MI, N, TE, R	1979–82	6	6	--
354	Unknown	PP, DS, MI, TE	1979	1	1	pH
355	Unknown	PP, DS, MI, N, TE, R	1980–83	8	8	pH, Sul (6), Fe (4), Mn (8)
356	Unknown	PP, DS, MI, N, TE	1975	1	1	Mn
357	Valley-fill deposits	PP, DS, MI, N, TE	1975	1	1	--
358	Valley-fill deposits	PP, DS, MI, N, TE	1975	2	2	Sul, Cu
359	Valley-fill deposits	PP, DS, MI, N, TE	1975	1	1	--
360	Unknown	PP	1975	1	1	--
361	Lewis Shale	PP, DS, MI, N, TE	1975	2	2	--
362	Alluvium, flood plain	PP	1975	1	1	--
363	Unknown	PP, DS, MI, N, TE	1975	1	1	Fe, Mn
364	Unknown	PP, DS, MI, N, TE	1975	2	2	--
365	Unknown	PP, DS, MI, N, TE	1975	1	1	--
366	Mesaverde Group	PP	1978	1	1	--
367	Lewis Shale	PP	1975	1	1	--
368	Mesaverde Group	PP	1978	1	1	--
369	Mancos Shale	PP, DS, MI, N, TE	1978	1	1	Fe
370	Unknown	PP	1975	1	1	--
371	Lewis Shale	PP, DS, MI, N, TE	1978–79	3	3	--
372	Lewis Shale	PP	1979	1	1	--
373	Lewis Shale	PP	1975	1	1	--
374	Lewis Shale	PP	1975	1	1	--
375	Lewis Shale	PP, DS, MI, N, TE	1975	1	1	NO_3+NO_2
376	Unknown	PP	1975	1	1	--

Appendix 5.   Description of selected groundwater sampling sites in the Upper Yampa River watershed, Colorado, with geologic unit description, type of water-quality data collected, period of water-quality record, number of samples collected, and constituents with exceedances of Colorado Department of Public Health and Environment water-quality standards for groundwater, 1975 through 1989 and 1998.—Continued

[No., number; CDPHE, Colorado Department of Public Health and Environment; USGS, U.S. Geological Survey; PP, physical properties; --, water-quality standard not exceeded; DS, dissolved solids; MI, major ions; N, nutrients; TE, trace elements; Fe, iron; Mn, manganese; NO_3+NO_2, nitrate plus nitrite; Zn, zinc; Sul, sulfate; Cd, cadmium; Ar, arsenic; R, radiochemical; Chl, chloride; Bo, boron; Pb, lead; SI, stable isotopes; OC, organic carbon; Fl, fluoride; Be, beryllium; Cu, copper; Se, selenium; Mo, molybdenum; CDOA, Colorado Department of Agriculture. The Upper Yampa River water-quality database is available at *http //rmgsc.cr.usgs.gov/cwqdr/Yampa/index.shtml*]

Site no. (see figure 15)	Site name in Upper Yampa River watershed water-quality database	Source of data	Site Identifier	Latitude	Longitude
377	SB00608817BBC1	USGS	402857107174201	40.482	−107.296
378	SB00608717BAD1	USGS	402858107102501	40.483	−107.174
379	SB00608514BBC1	USGS	402902106470401	40.484	−106.785
380	SB00608717BAA1	USGS	402902107101801	40.484	−107.172
381	SB00608614ABA1	USGS	402904106595701	40.484	−107.000
382	SB00608011AAA1	USGS	402906107040801	40.485	−107.069
383	SB00608912DDD1	USGS	402906107185401	40.485	−107.316
384	SB00608509DDD1 USGS 402907106550300	USGS	402907106550300	40.485	−106.918
385	SB00608615BAA1	USGS	402908107011901	40.486	−107.023
386	SB00608708CCD1	USGS	402908107103701	40.486	−107.178
387	SB00608609DCD1	USGS	402912107020901	40.487	−107.036
388	SB00608707DCC1	USGS	402912107112401	40.487	−107.191
389	SB00608611DDD1	USGS	402913107051501	40.487	−107.088
390	SB00608708CCD2	USGS	402913107103701	40.487	−107.178
391	SB00608811DDC1	USGS	402913107132101	40.487	−107.223
392	SB00508612DBB1 USGS 402914106585701	USGS	402914106585701	40.487	−106.983
393	SB00608712DDA	USGS	402916107061601	40.488	−107.105
394	SB00608808DCB1	USGS	402917107170101	40.488	−107.284
395	SB00608509DAD1	USGS	402920106550701	40.489	−106.919
396	SB00608809CBD1	USGS	402920107154301	40.489	−107.263
397	SB00608708CBD1	USGS	402921107104001	40.489	−107.178
398	SB00608708CBD2	USGS	402922107104001	40.489	−107.178
399	SB00608511CAD1	USGS	402924106532301	40.490	−106.890
400	SB00608707DBC1	USGS	402924107112001	40.490	−107.190
401	SB00608707DBC2	USGS	402924107112201	40.490	−107.190
402	SB00608609DAB1	USGS	402929107015901	40.491	−107.034
403	SB00608707DAB1	USGS	402929107110301	40.491	−107.185
404	SB00608807DBB1	USGS	402929107181701	40.491	−107.305
405	SB00608707DAA1	USGS	402930107105101	40.492	−107.181
406	SB00608707DAA2	USGS	402930107105201	40.492	−107.182
407	SB00608911DBB1	USGS	402930107203001	40.492	−107.342
408	SB00608508CAB1	USGS	402931106565501	40.492	−106.949
409	SB00608609DBA1	USGS	402931107020901	40.492	−107.036
410	SB00608911ACC1	USGS	402931107203001	40.492	−107.342
411	SB00608707ADC1	USGS	402932107110101	40.492	−107.184
412	SB00608707DBB1	USGS	402932107112001	40.492	−107.190
413	SB00608808CAB	USGS	402932107171801	40.492	−107.289
414	SB00608707ADC2	USGS	402933107110201	40.492	−107.185
415	SB00608707ADC3	USGS	402933107110701	40.492	−107.186
416	SB00608707ADC4	USGS	402933107110702	40.492	−107.186
417	SB00608707ACD1	USGS	402933107111301	40.492	−107.188
418	SB00608707ACC1	USGS	402933107112201	40.492	−107.190
419	SB00608707ACD2	USGS	402934107111301	40.493	−107.188
420	SB00608707ADD1	USGS	402935107105301	40.493	−107.182
421	SB00608707ACD3	USGS	402937107111401	40.494	−107.188
422	SB00608809ACD1	USGS	402938107155201	40.494	−107.265
423	SB00608912ADA1	USGS	402938107185201	40.494	−107.315
424	SB00608512BCA1	USGS	402941106523101	40.495	−106.876
425	SB00608809ADA1	USGS	402942107152501	40.495	−107.258
426	SB00608809ACA1	USGS	402942107154401	40.495	−107.263
427	SB00608510ACA1	USGS	402944106541301	40.496	−106.904

Appendix 5. Description of selected groundwater sampling sites in the Upper Yampa River watershed, Colorado, with geologic unit description, type of water-quality data collected, period of water-quality record, number of samples collected, and constituents with exceedances of Colorado Department of Public Health and Environment water-quality standards for groundwater, 1975 through 1989 and 1998.—Continued

[No., number; CDPHE, Colorado Department of Public Health and Environment; USGS, U.S. Geological Survey; PP, physical properties; --, water-quality standard not exceeded; DS, dissolved solids; MI, major ions; N, nutrients; TE, trace elements; Fe, iron; Mn, manganese; NO_3+NO_2, nitrate plus nitrite; Zn, zinc; Sul, sulfate; Cd, cadmium; Ar, arsenic; R, radiochemical; Chl, chloride; Bo, boron; Pb, lead; SI, stable isotopes; OC, organic carbon; Fl, fluoride; Be, beryllium; Cu, copper; Se, selenium; Mo, molybdenum; CDOA, Colorado Department of Agriculture. The Upper Yampa River water-quality database is available at *http //rmgsc.cr.usgs.gov/cwqdr/Yampa/index.shtml*]

Site no. (see figure 15)	Geologic unit description	Type of water-quality data collected[1]	Period of water-quality record (calendar year)	No. of sample days	No. of samples[2]	Constituent with exceedance of CDPHE water-quality standard[3]
377	Alluvium, flood plain	PP	1975	1	1	--
378	Alluvium, terrace	PP, DS, MI, N, TE	1979	5	5	Bo (3), Mn (2), Se
379	Alluvium, terrace	PP	1978	1	1	--
380	Alluvium, terrace	PP, DS, MI, N, TE	1978–79	6	6	Sul (2), Fe, Mn (3)
381	Alluvium, flood plain	PP	1978	1	1	pH
382	Alluvium, flood plain	PP	1975	1	1	--
383	Alluvium, flood plain	PP	1975	1	1	--
384	Unknown	PP	1975	1	1	--
385	Alluvium, flood plain	PP	1975	1	1	--
386	Alluvium, terrace	PP, DS, MI, N, TE	1978–79	8	8	Bo, Mn
387	Alluvium, flood plain	PP	1978	1	1	pH
388	Alluvium, terrace	PP, DS, MI, N, TE	1978–79	6	6	NO_3+NO_2 (3), Mn
389	Alluvium, flood plain	PP	1975	1	1	--
390	Alluvium, terrace	PP, DS, MI, N, TE	1978–79	6	6	Bo (3), Mn (2)
391	Lewis Shale	PP	1978	1	1	pH
392	Unknown	PP	1975	1	1	--
393	Mesaverde Group	PP, DS, MI, N, TE	1975	1	1	--
394	Alluvium, flood plain	PP	1975	1	1	--
395	Mancos Shale	PP	1978	1	1	--
396	Alluvium, flood plain	PP	1978	1	1	--
397	Alluvium, terrace	PP, DS, MI, N, TE	1978–79	6	6	Sul, Bo (4), Mn
398	Alluvium, terrace	PP, DS, MI, N, TE	1978–79	5	5	Sul, Bo (4), Mn (3)
399	Browns Park Formation	PP, DS, MI, N, TE	1978	1	1	pH
400	Alluvium, terrace	PP, DS, MI, N, TE	1978–79	5	5	Bo (4), Mn (2)
401	Alluvium, terrace	PP, DS, MI, N, TE	1978–79	6	6	Sul, Bo (4), Mn
402	Mesaverde Group	PP	1978	1	1	pH
403	Alluvium, terrace	PP, DS, MI, N, TE	1978–79	4	4	Sul (2), Bo (2), Mn (2)
404	Alluvium, flood plain	PP	1975	1	1	--
405	Alluvium, terrace	PP, DS, MI, N, TE	1978–79	5	5	Sul (2), Bo (4), Mn (3)
406	Alluvium, terrace	PP, DS, MI, N, TE	1978–79	6	6	Sul (2), Bo (4), Mn
407	Upper Cretaceous series	PP	1978	1	1	--
408	Alluvium, flood plain	PP	1975	1	1	--
409	Alluvium, flood plain	PP	1975	1	1	--
410	Alluvium, flood plain	PP	1975	1	1	--
411	Alluvium, terrace	PP, DS, MI, N, TE	1978–79	5	5	Bo (3), Mn (2)
412	Alluvium, terrace	PP, DS, MI, N, TE	1978–79	5	5	Sul (2), Bo (4)
413	Alluvium, flood plain	PP, DS, MI, N, TE	1975	2	2	Sul, Fe, Mn
414	Alluvium, terrace	PP, DS, MI, N, TE	1978–79	5	5	Bo (2), Mn (2)
415	Alluvium, terrace	PP, DS, MI, N, TE	1978–79	5	5	Sul (2), Bo (3), Mn (3)
416	Alluvium, terrace	PP, DS, MI, N, TE	1978–79	5	5	Sul (2), Bo (3), Fe, Mn (4)
417	Alluvium, terrace	PP, DS, MI, N, TE	1978–79	5	5	Sul (2), Bo (4), Mn (2)
418	Alluvium, terrace	PP, DS, MI, N, TE	1978–79	6	6	Sul (2), Bo (3)
419	Alluvium, terrace	PP, DS, MI, N, TE	1978–79	6	6	Sul (2), Bo (4), Mn (2)
420	Alluvium, terrace	PP, DS, MI, N, TE	1978–79	6	6	Bo (2)
421	Alluvium, terrace	PP, DS, MI, N, TE	1978–79	5	5	Sul (2), Bo (4), Mn
422	Alluvium, flood plain	PP	1978	1	1	--
423	Alluvium, flood plain	PP	1975	1	1	--
424	Mancos Shale	PP	1978	1	1	--
425	Alluvium, flood plain	PP	1978	1	1	--
426	Alluvium, flood plain	PP	1978	1	1	--
427	Unknown	PP	1975	1	1	--

Appendix 5.   Description of selected groundwater sampling sites in the Upper Yampa River watershed, Colorado, with geologic unit description, type of water-quality data collected, period of water-quality record, number of samples collected, and constituents with exceedances of Colorado Department of Public Health and Environment water-quality standards for groundwater, 1975 through 1989 and 1998.—Continued

[No., number; CDPHE, Colorado Department of Public Health and Environment; USGS, U.S. Geological Survey; PP, physical properties; --, water-quality standard not exceeded; DS, dissolved solids; MI, major ions; N, nutrients; TE, trace elements; Fe, iron; Mn, manganese; NO_3+NO_2, nitrate plus nitrite; Zn, zinc; Sul, sulfate; Cd, cadmium; Ar, arsenic; R, radiochemical; Chl, chloride; Bo, boron; Pb, lead; SI, stable isotopes; OC, organic carbon; Fl, fluoride; Be, beryllium; Cu, copper; Se, selenium; Mo, molybdenum; CDOA, Colorado Department of Agriculture. The Upper Yampa River water-quality database is available at *http://rmgsc.cr.usgs.gov/cwqdr/Yampa/index.shtml*]

Site no. (see figure 15)	Site name in Upper Yampa River watershed water-quality database	Source of data	Site Identifier	Latitude	Longitude
428	SB00608507BCB1	USGS	402946106582101	40.496	−106.973
429	SB00608911BAD1	USGS	402948107203801	40.497	−107.345
430	SB00608509BBC1	USGS	402949106491801	40.497	−106.822
431	SB00608811BAC1	USGS	402949107135201	40.497	−107.232
432	SB00608708BBA1	USGS	402954107103501	40.498	−107.177
433	SB00608708BAB1	USGS	402957107103301	40.499	−107.176
434	SB00608802DDD1	USGS	402959107131101	40.500	−107.220
435	SB00608904CDD1	USGS	402959107222701	40.500	−107.375
436	HS-13	USGS	402960107112401	40.500	−107.191
437	SB00608704DCA1	USGS	403003107224201	40.501	−107.379
438	SB00608503CCC	USGS	403004106550001	40.501	−106.917
439	SB00608406CCA1	USGS	403005106512101	40.501	−106.856
440	SB00708503CAD1	USGS	403012106542201	40.503	−106.907
441	SB00608804DBD1	USGS	403014107154701	40.504	−107.264
442	SB00608505DAD1	USGS	403015106561001	40.504	−106.937
443	SB00608501CBC1	USGS	403017106523701	40.505	−106.878
444	SB00608503ADC1	USGS	403025106540101	40.507	−106.901
445	SB00708506CAA1	USGS	403025106575201	40.507	−106.965
446	SB00608705BDD1	USGS	403031107101701	40.509	−107.172
447	SB00608801ADC1 USGS 403031107121601	USGS	403031107121601	40.509	−107.205
448	SB00608801ADC1 USGS 403032107115701	USGS	403032107115701	40.509	−107.205
449	SB00608703ACA1	USGS	403035107075401	40.510	−107.132
450	SB00608904BAB1	USGS	403041107230601	40.511	−107.386
451	SB00608803BAA1	USGS	403047107150701	40.513	−107.253
452	SB00708433CBC1	USGS	403106106492001	40.518	−106.823
453	SB00708731BCB1	USGS	403125107115701	40.524	−107.200
454	SB00708433ABD1	USGS	403127106483501	40.524	−106.810
455	SB00708532BCB1	USGS	403129106570801	40.525	−106.953
456	SB00708827CBC1 USGS 403132107152501	USGS	403132107152501	40.526	−107.258
457	SB00708433BBB1	USGS	403136106491401	40.527	−106.821
458	SB00709025DDD	USGS	403143107254501	40.529	−107.430
459	SB00709025CAC1	USGS	403156107262501	40.532	−107.441
460	SB00708428CAD1	USGS	403202106485601	40.534	−106.816
461	SB00708428CBB1	USGS	403205106492001	40.535	−106.823
462	SB00708428CBA1	USGS	403207106490601	40.535	−106.819
463	SB00708525ACC1	USGS	403210106520801	40.536	−106.869
464	SB00708426BCD1	USGS	403211106465401	40.536	−106.782
465	SB00708428ACD1	USGS	403211106484001	40.536	−106.812
466	UNKNOWN	USGS	403218106532601	40.538	−106.891
467	SB00708526ACB1	USGS	403219106531101	40.539	−106.887
468	SB00709025BBD1	USGS	403219107263800	40.539	−107.445
469	SB00708525BAD1	USGS	403224106521501	40.540	−106.871
470	SB00708924DDD1	USGS	403233107185801	40.542	−107.317
471	SB00708920CCC1	USGS	403235107243001	40.543	−107.409
472	SB00708525ABD1	USGS	403242106520301	40.545	−106.868
473	SB00708920DCB	USGS	403245107235801	40.546	−107.400
474	SB00708523DBB USGS 403258106531300	USGS	403258106531300	40.549	−106.888
475	SB00708924ADD USGS 403303107185600	USGS	403303107185600	40.551	−107.316
476	SB00708829CCD1	USGS	403318107172501	40.555	−107.291
477	SB00708614DAB1	USGS	403351106594201	40.564	−106.996

Appendix 5.   Description of selected groundwater sampling sites in the Upper Yampa River watershed, Colorado, with geologic unit description, type of water-quality data collected, period of water-quality record, number of samples collected, and constituents with exceedances of Colorado Department of Public Health and Environment water-quality standards for groundwater, 1975 through 1989 and 1998.—Continued

[No., number; CDPHE, Colorado Department of Public Health and Environment; USGS, U.S. Geological Survey; PP, physical properties; --, water-quality standard not exceeded; DS, dissolved solids; MI, major ions; N, nutrients; TE, trace elements; Fe, iron; Mn, manganese; NO_3+NO_2, nitrate plus nitrite; Zn, zinc; Sul, sulfate; Cd, cadmium; Ar, arsenic; R, radiochemical; Chl, chloride; Bo, boron; Pb, lead; SI, stable isotopes; OC, organic carbon; Fl, fluoride; Be, beryllium; Cu, copper; Se, selenium; Mo, molybdenum; CDOA, Colorado Department of Agriculture. The Upper Yampa River water-quality database is available at *http //rmgsc.cr.usgs.gov/cwqdr/Yampa/index.shtml*]

Site no. (see figure 15)	Geologic unit description	Type of water-quality data collected[1]	Period of water-quality record (calendar year)	No. of sample days	No. of samples[2]	Constituent with exceedance of CDPHE water-quality standard[3]
428	Alluvium, flood plain	PP	1975	1	1	--
429	Unknown	PP	1975	1	1	--
430	Mancos Shale	PP	1978	1	1	--
431	Unknown	PP	1975	1	1	--
432	Alluvium, flood plain	PP	1975	1	1	--
433	Alluvium, flood plain	PP, DS, MI, N, TE	1978	1	1	Fe, Mn
434	Alluvium, flood plain	PP	1975	1	1	--
435	Alluvium, flood plain	PP	1975	1	1	--
436	Unknown	PP, DS, MI, N, TE	1978	1	1	Bo
437	Alluvium, flood plain	PP	1975	1	1	--
438	Alluvium, flood plain	PP	1975	2	2	--
439	Alluvium, flood plain	PP	1978	1	1	--
440	Alluvium, flood plain	PP, DS, MI, N, TE	1975	2	2	Mn
441	Alluvium, flood plain	PP	1975	1	1	--
442	Unknown	PP	1975	1	1	--
443	Alluvium, flood plain	PP, DS, MI, N, TE	1975	1	1	Mn
444	Unknown	PP	1975	1	1	--
445	Alluvium, flood plain	PP, DS, MI, N, TE	1975	2	2	--
446	Alluvium, flood plain	PP	1975	1	1	--
447	Alluvium, flood plain	PP	1978	1	1	--
448	Valley-fill deposits	PP, DS, MI, N, TE	1975, 1978	3	3	Fe
449	Unknown	PP	1975	1	1	--
450	Alluvium, flood plain	PP	1975	1	1	--
451	Alluvium, flood plain	PP	1975	1	1	--
452	Alluvium, flood plain	PP, DS, MI, N, TE	1978	1	1	pH
453	Alluvium, flood plain	PP	1975	1	1	--
454	Precambrian Erathem	PP	1978	1	1	--
455	Mancos Shale	PP, DS, MI, N, TE	1975	2	2	pH, Chl, Fl
456	Lewis Shale	PP, DS, MI, N, TE	1975	1	1	Sul, NO_3+NO_2
457	Browns Park Formation	PP	1978	1	1	--
458	Unknown	PP	1975	1	1	--
459	Alluvium, flood plain	PP	1975	1	1	--
460	Alluvium, flood plain	PP	1978	1	1	--
461	Browns Park Formation	PP, DS, MI, N, TE	1978	1	1	--
462	Browns Park Formation	PP, DS, MI, N, TE	1978	1	1	Fl
463	Alluvium, flood plain	PP	1978	1	1	--
464	Precambrian Erathem	PP, DS, MI, N, TE	1978	1	1	Fe, Mn
465	Alluvium, flood plain	PP	1978	1	1	--
466	Unknown	PP, DS, MI, N, TE	1978	1	1	Sul
467	Mancos Shale	PP, DS, MI, N, TE	1978	1	1	Fe, Mn, Mo
468	Unknown	PP	1975	1	1	--
469	Mancos Shale	PP	1978	1	1	--
470	Upper Cretaceous series	PP	1978	1	1	--
471	Eocene series	PP, DS, MI, N, TE	1975	2	2	--
472	Mancos Shale	PP	1978	1	1	--
473	Alluvium, flood plain	PP	1975	1	1	--
474	Mancos Shale	PP, DS, MI, N, TE	1975	1	1	Chl, Fl
475	Upper Cretaceous series	PP, DS, MI, N, TE	1975	1	1	Mn
476	Unknown	PP	1975	1	1	--
477	Mancos Shale	PP	1978	1	1	--

Appendix 5.   Description of selected groundwater sampling sites in the Upper Yampa River watershed, Colorado, with geologic unit description, type of water-quality data collected, period of water-quality record, number of samples collected, and constituents with exceedances of Colorado Department of Public Health and Environment water-quality standards for groundwater, 1975 through 1989 and 1998.—Continued

[No., number; CDPHE, Colorado Department of Public Health and Environment; USGS, U.S. Geological Survey; PP, physical properties; --, water-quality standard not exceeded; DS, dissolved solids; MI, major ions; N, nutrients; TE, trace elements; Fe, iron; Mn, manganese; NO_3+NO_2, nitrate plus nitrite; Zn, zinc; Sul, sulfate; Cd, cadmium; Ar, arsenic; R, radiochemical; Chl, chloride; Bo, boron; Pb, lead; SI, stable isotopes; OC, organic carbon; Fl, fluoride; Be, beryllium; Cu, copper; Se, selenium; Mo, molybdenum; CDOA, Colorado Department of Agriculture. The Upper Yampa River water-quality database is available at *http //rmgsc.cr.usgs.gov/cwqdr/Yampa/index.shtml*]

Site no. (see figure 15)	Site name in Upper Yampa River watershed water-quality database	Source of data	Site Identifier	Latitude	Longitude
478	SB00708912AAA1	USGS	403504107182701	40.584	−107.308
479	SB007086006DCD1	USGS	403516107042501	40.588	−107.074
480	SB00708505CAA1	USGS	403536106564701	40.593	−106.947
481	SB00708829BCC	USGS	403537107162901	40.594	−107.275
482	SB00808933ABD IVAN KAWCAK	USGS	403646107223700	40.613	−107.378
483	SB00808529DDD1	USGS	403702106591901	40.617	−106.989
484	SB00808629DAD1 USGS 403711107030101	USGS	403711107030101	40.620	−107.051
485	SB00808629DAD1 USGS 403712107030201	USGS	403712107030201	40.620	−107.051
486	SB00808623CCD1	USGS	403754107002301	40.632	−107.007
487	SB00808520DAC1	USGS	403810106563001	40.636	−106.942
488	SB00808923ADA1	USGS	403820107200001	40.639	−107.334
489	SB00808622BCD1	USGS	403821107013501	40.639	−107.027
490	SB00808519BBA1	USGS	403843106581501	40.645	−106.971
491	SB00808613DDC1	USGS	403843106583501	40.645	−106.977
492	SB00808517DCA1	USGS	403855106563401	40.649	−106.943
493	SB00808517DBC1	USGS	403903106564301	40.651	−106.946
494	SB00808517BDD W C WINTER	USGS	403912106565000	40.653	−106.948
495	SB00908527BBC1	USGS	404259106542301	40.716	−106.907
496	SB00908527BBB1	USGS	404302106545201	40.717	−106.915
497	SB00908522CDD	USGS	404309106543001	40.719	−106.909
498	SB00908523CCB1 USGS 404318106534201	USGS	404318106534201	40.722	−106.896
499	SB00908513CDC1	USGS	404402106521001	40.734	−106.870
500	SB00908513DAD1	USGS	404414106515101	40.737	−106.865
501	SB00908518ABC1	USGS	404446106574501	40.746	−106.963
502	WS-027CDOA	CDOA	WS-027	40.19	−106.91
503	WS-029CDOA	CDOA	WS-029	40.48	−107.02
504	WS-031CDOA	CDOA	WS-031	40.5	−107.18
505	WS-032CDOA	CDOA	WS-032	40.49	−107.3
506	WS-033CDOA	CDOA	WS-033	40.56	−106.89
507	WS-034CDOA	CDOA	WS-034	40.19	−106.92

Appendix 5. Description of selected groundwater sampling sites in the Upper Yampa River watershed, Colorado, with geologic unit description, type of water-quality data collected, period of water-quality record, number of samples collected, and constituents with exceedances of Colorado Department of Public Health and Environment water-quality standards for groundwater, 1975 through 1989 and 1998.—Continued

[No., number; CDPHE, Colorado Department of Public Health and Environment; USGS, U.S. Geological Survey; PP, physical properties; --, water-quality standard not exceeded; DS, dissolved solids; MI, major ions; N, nutrients; TE, trace elements; Fe, iron; Mn, manganese; NO_3+NO_2, nitrate plus nitrite; Zn, zinc; Sul, sulfate; Cd, cadmium; Ar, arsenic; R, radiochemical; Chl, chloride; Bo, boron; Pb, lead; SI, stable isotopes; OC, organic carbon; Fl, fluoride; Be, beryllium; Cu, copper; Se, selenium; Mo, molybdenum; CDOA, Colorado Department of Agriculture. The Upper Yampa River water-quality database is available at *http //rmgsc.cr.usgs.gov/cwqdr/Yampa/index.shtml*]

Site no. (see figure 15)	Geologic unit description	Type of water-quality data collected[1]	Period of water-quality record (calendar year)	No. of sample days	No. of samples[2]	Constituent with exceedance of CDPHE water-quality standard[3]
478	Unknown	PP	1975	1	1	--
479	Mancos Shale	PP, DS, MI, N, TE	1975, 1978	3	3	Zn (2)
480	Mancos Shale	PP, DS, MI, N, TE	1978	1	1	Sul, Chl, Fe, Mn
481	Unknown	PP	1975	1	1	--
482	Fort Union Formation	PP, DS, MI, N, TE	1975	1	1	NO_3+NO_2, Fe
483	Alluvium, flood plain	PP	1978	1	1	--
484	Alluvium, flood plain	PP	1975	1	1	--
485	Alluvium, flood plain	PP, DS, MI, N, TE	1978	1	1	--
486	Alluvium, flood plain	PP, DS, MI, N, TE	1975	2	2	--
487	Mancos Shale	PP, DS, MI, N, TE	1978	1	1	Bo
488	Unknown	PP	1975	1	1	--
489	Alluvium, flood plain	PP	1975	1	1	--
490	Mancos Shale	PP	1978	1	1	--
491	Mancos Shale	PP	1978	1	1	--
492	Mancos Shale	PP	1978	1	1	--
493	Alluvium, flood plain	PP	1978	1	1	pH
494	Valley-fill deposits	PP, DS, MI, N, TE	1975	1	1	pH, Fe
495	Mancos Shale	PP	1978	1	1	pH
496	Alluvium, flood plain	PP	1978	1	1	pH
497	Alluvium, flood plain	PP	1978	1	1	--
498	Alluvium, flood plain	PP, DS, MI, N, TE	1978	2	2	pH (2)
499	Alluvium, flood plain	PP	1978	1	1	pH
500	Alluvium, flood plain	PP	1978	1	1	--
501	Browns Park Formation	PP, DS, MI, N, TE	1978	1	1	Mn
502	Unknown	PP, DS, MI, N	1998	1	1	--
503	Unknown	PP, DS, MI, N, TE	1998	1	1	--
504	Unknown	PP, DS, MI, N, TE	1998	1	1	--
505	Unknown	PP, DS, MI, N, TE	1998	1	1	--
506	Unknown	PP, DS, MI, N, TE	1998	1	1	--
507	Unknown	PP, DS, MI, N	1998	1	1	--

[1]When multiple samples were collected at a site, all types of water-quality data listed may not be available for each sample.

[2]Count does not include samples with water-level measurements only.

[3]Number in parentheses is number of exceedances. Except for pH, all standards are for dissolved water samples.

[4]Sample collection did not occur in every year of the period of record.

Appendix 6.   Description of stream sites in the Upper Yampa River watershed, Colorado, that have macroinvertebrate data, and period of water-quality record and number of sample days, 1975 through 2008.

[No., number; USGS, U.S. Geological Survey; CDPHE, Colorado Department of Public Health and Environment; CSS, City of Steamboat Springs. Subwatershed definitions: Yampa River subwatershed 1, Yampa River and tributaries upstream from Chuck Lewis State Wildlife Area; Yampa River subwatershed 2, Yampa River and tributaries from Chuck Lewis State Wildlife Area to Elk River confluence; Elk River subwatershed, Elk River and tributaries; Yampa River subwatershed 3, Yampa River and tributaries from Elk River confluence to Town of Hayden; Yampa River subwatershed 4, Yampa River and tributaries from Town of Hayden to Elkhead Creek confluence; Elkhead Creek subwatershed, Elkhead Creek and tributaries. Sites with the same site number are considered to be at the same location]

Site no. (see figure 18)	Site name	Source of data	Site identifier
6	BEAR RIVER NEAR TOPONAS, CO	USGS	09236000
7	BEAVER CREEK NEAR HAHNS PEAK, CO.	USGS	404610106545600
10	BURGESS CK NEAR MOUTH @ HWY 40	CDPHE	12893
180	Bushy Creek above Cty Rd 16	CDPHE	12885
22	DRY CK @ HAYDEN	CDPHE	12852
27	ELK R. NEAR MOUTH @ CR44	CDPHE	12860
30	ELK RIVER AT CLARK, CO.	USGS	09241000
33	ELK RIVER NEAR MILNER, CO.	USGS	09242500
38	ELKHEAD CREEK	CDPHE	WCOP99-0512
39	ELKHEAD CREEK	CDPHE	WCOP99-0565
40	ELKHEAD CREEK ABOVE ELKHEAD RESERVOIR, CO.	USGS	403530107191300
181	Elkhead Creek above First Creek	CDPHE	12849A
182	Elkhead Creek above Rippy property	CDPHE	12846A
183	Elkhead Creek at County Rd. 76	CDPHE	12843
184	Elkhead Creek at Rippy property	CDPHE	12846
185	Elkhead Creek below Elkhead Reservoir	CDPHE	12843A
43	ELKHEAD CREEK NEAR CRAIG, CO	USGS	09246500
43	ELKHEAD CK NR CRAIG @ HWY 40	CDPHE	12840
44	ELKHEAD CREEK NEAR ELKHEAD, CO.	USGS	09245000
186	First Creek east of USFS/State boundary	CDPHE	12849
54	FISH CK NEAR MOUTH @ HWY 40	CDPHE	12870
58	GRASSY CREEK AT GRASSY GAP, CO.	USGS	402330107082000
59	GRASSY CREEK NEAR MOUNT HARRIS, CO.	USGS	09244300
59	GRASSY CK @ RD. 27A	CDPHE	12853
187	Grassy Creek at Rd 27 STL	CDPHE	12853A
69	MAD CK @ CHRISTINA SWA	CDPHE	12863
81	NORTH FORK WALTON CREEK NR RABBIT EARS PASS, CO.	USGS	09238300
83	OAK CK D/S TOWN OF OAK CREEK @ CR 27	CDPHE	12892
85	OAK CREEK AB OAK CREEK DRAIN NEAR OAK CREEK, CO.	USGS	401725106575600
93	OAK CREEK NEAR OAK CREEK, CO.	USGS	09238000
94	OAK CK @ 22 RD ABV YAMPA R	CDPHE	12891
95	PHILLIPS CREEK NEAR YAMPA, CO.	USGS	400759106532500
98	SAGE CK @ RD. 27	CDPHE	12851
188	Sage Creek in canyon on Rd 37	CDPHE	12851B
101	SAGE CREEK NEAR HAYDEN, CO.	USGS	402918107094400
102	SAGE CREEK NEAR MOUNT HARRIS, CO.	USGS	402522107134100
115	TROUT CK NR. MOUTH	CDPHE	12876
116	TROUT CREEK ABOVE FOIDEL CREEK NEAR MILNER, CO.	USGS	402416106580800
121	TROUT CREEK NEAR PHIPPSBURG, CO.	USGS	09243000
122	WALTON CR. NEAR MOUTH @ HWY 40	CDPHE	12894
189	Wolf Creek at 52 Rd	CDPHE	12855
128	WOLF CREEK NEAR HAYDEN, CO.	USGS	402832107080200
133	YMP-7 [Yampa River 100m above James Brown Bridge]	CSS[1]	YMP-7
136	YAMPA R. @ CR 14 FISHING ACCESS	CDPHE	12806D
137	YAMPA R. ABV. PHIPPSBURG	CDPHE	12814
138	YAMPA R. BLW STAGECOACH RES.	CDPHE	12808
139	YAMPA R. D/S STAGECOACH RES. DAM	CDPHE	12808P
140	YAMPA R. N. OF HAYDEN @ CALIFORNIA PARK RD	CDPHE	12802
141	YAMPA R. NR MOUNT HARRIS BLW HWY 40 BRIDGE	CDPHE	12805
142	YAMPA R. U/S LAKE CATAMOUNT @ CR18	CDPHE	12807

Appendix 6.   Description of stream sites in the Upper Yampa River watershed, Colorado, that have macroinvertebrate data, and period of water-quality record and number of sample days, 1975 through 2008.—Continued

[No., number; USGS, U.S. Geological Survey; CDPHE, Colorado Department of Public Health and Environment; CSS, City of Steamboat Springs. Subwatershed definitions: Yampa River subwatershed 1, Yampa River and tributaries upstream from Chuck Lewis State Wildlife Area; Yampa River subwatershed 2, Yampa River and tributaries from Chuck Lewis State Wildlife Area to Elk River confluence; Elk River subwatershed, Elk River and tributaries; Yampa River subwatershed 3, Yampa River and tributaries from Elk River confluence to Town of Hayden; Yampa River subwatershed 4, Yampa River and tributaries from Town of Hayden to Elkhead Creek confluence; Elkhead Creek subwatershed, Elkhead Creek and tributaries. Sites with the same site number are considered to be at the same location]

Site no. (see figure 18)	Latitude	Longitude	Subwatershed	Period of water-quality record (month/year)	No. of sample days
6	40.044	−107.072	Yampa River subwatershed 1	8/1975	1
7	40.769	−106.916	Elk River subwatershed	9/1975	1
10	40.452	−106.810	Yampa River subwatershed 2	4/2001, 8/2001	2
180	40.201	−106.826	Yampa River subwatershed 2	9/2006	1
22	40.492	−107.265	Yampa River subwatershed 4	4/2003	1
27	40.546	−106.909	Elk River subwatershed	4/2001, 8/2001	2
30	40.717	−106.916	Elk River subwatershed	8/1975	1
33	40.515	−106.954	Elk River subwatershed	8/1975	1
38	40.660	−107.291	Elkhead Creek subwatershed	8/2000	1
39	40.620	−107.271	Elkhead Creek subwatershed	8/2001, 7/2008	2
40	40.592	−107.321	Elkhead Creek subwatershed	9/1975	1
181	40.755	−107.133	Elkhead Creek subwatershed	7/2008	1
182	40.676	−107.271	Elkhead Creek subwatershed	7/2008	1
183	40.592	−107.321	Elkhead Creek subwatershed	7/2008	1
184	40.669	−107.285	Elkhead Creek subwatershed	7/2008	1
185	40.539	−107.411	Elkhead Creek subwatershed	7/2008	1
43	40.531	−107.436	Elkhead Creek subwatershed	9/1975	1
43	40.531	−107.436	Elkhead Creek subwatershed	8/2001, 7/2008	2
44	40.670	−107.285	Elkhead Creek subwatershed	9/1975	1
186	40.731	−107.147	Elkhead Creek subwatershed	8/2005	1
54	40.467	−106.825	Yampa River subwatershed 2	4/2001, 8/2001	2
58	40.392	−107.139	Yampa River subwatershed 3	9/1975	1
59	40.447	−107.146	Yampa River subwatershed 3	9/1975	1
59	40.447	−107.146	Yampa River subwatershed 3	4/2003	1
187	40.390	−107.148	Yampa River subwatershed 3	4/2003	1
69	40.565	−106.889	Elk River subwatershed	10/1997, 9/1998, 7/2008	3
81	40.396	−106.650	Yampa River subwatershed 2	8/1975	1
83	40.276	−106.964	Yampa River subwatershed 1	4/2001, 8/2001	2
85	40.290	−106.966	Yampa River subwatershed 1	8/1975	1
93	40.244	−107.015	Yampa River subwatershed 1	8/1975	1
94	40.399	−106.842	Yampa River subwatershed 1	4/2001, 8/2001	2
95	40.133	−106.891	Yampa River subwatershed 1	8/1975	1
98	40.484	−107.170	Yampa River subwatershed 3	4/2003	1
188	40.404	−107.224	Yampa River subwatershed 3	7/2008	1
101	40.488	−107.163	Yampa River subwatershed 3	9/1975	1
102	40.423	−107.229	Yampa River subwatershed 3	9/1975	1
115	40.460	−106.989	Yampa River subwatershed 3	9/1998	1
116	40.404	−106.969	Yampa River subwatershed 3	8/1975	1
121	40.151	−107.132	Yampa River subwatershed 3	8/1975	1
122	40.270	−106.816	Yampa River subwatershed 2	4/2001, 8/2001	2
189	40.547	−107.113	Yampa River subwatershed 4	4/2003	1
128	40.476	−107.134	Yampa River subwatershed 4	9/1975	1
133	40.496	−106.857	Yampa River subwatershed 2	9/2005, 9/2007, 9/2008	[2]3
136	40.475	−106.824	Yampa River subwatershed 1	4/2001, 8/2001	2
137	40.227	−106.941	Yampa River subwatershed 1	9/1998, 4/2001, 8/2001	3
138	40.287	−106.829	Yampa River subwatershed 1	8/1998	1
139	40.288	−106.827	Yampa River subwatershed 1	4/2001, 8/2001	2
140	40.502	−107.264	Yampa River subwatershed 3	4/2001, 8/2001	2
141	40.488	−107.158	Yampa River subwatershed 3	4/2001, 8/2001	2
142	40.341	−106.808	Yampa River subwatershed 1	4/2001, 8/2001	2

Appendix 6. Description of stream sites in the Upper Yampa River watershed, Colorado, that have macroinvertebrate data, and period of water-quality record and number of sample days, 1975 through 2008.—Continued

[No , number; USGS, U.S. Geological Survey; CDPHE, Colorado Department of Public Health and Environment; CSS, City of Steamboat Springs. Subwatershed definitions: Yampa River subwatershed 1, Yampa River and tributaries upstream from Chuck Lewis State Wildlife Area; Yampa River subwatershed 2, Yampa River and tributaries from Chuck Lewis State Wildlife Area to Elk River confluence; Elk River subwatershed, Elk River and tributaries; Yampa River subwatershed 3, Yampa River and tributaries from Elk River confluence to Town of Hayden; Yampa River subwatershed 4, Yampa River and tributaries from Town of Hayden to Elkhead Creek confluence; Elkhead Creek subwatershed, Elkhead Creek and tributaries. Sites with the same site number are considered to be at the same location]

Site no. (see figure 18)	Site name	Source of data	Site identifier
143	YAMPA RIVER AB OAK CREEK NR STEAMBOAT SPGS, CO.	USGS	402356106500000
143	YAMPA R. ABV OAK CREEK	CDPHE	12811
145	YAMPA RIVER ABOVE ELK RIVER NEAR MILNER, CO.	USGS	402932106564900
147	YAMPA R. U/S STAGECOACH RES @ CR16	CDPHE	12809
150	YAMPA RIVER AT HAYDEN, CO.	USGS	403006107154800
151	YAMPA RIVER AT MILNER	CDPHE	000038
152	YAMPA RIVER AT PHIPPSBURG, CO.	USGS	401418106562200
153	YAMPA RIVER AT STEAMBOAT SPRINGS, CO	USGS	09239500
153	YAMPA R. @ 5TH ST. BRIDGE IN STEAMBOAT	CDPHE	12806
154	YAMPA RIVER BELOW DIVERSION, NEAR HAYDEN, CO.	USGS	09244410
155	YAMPA RIVER BELOW HAYDEN, CO.	USGS	402930107174200
162	YAMPA R. BLW YAMPA @ CR21	CDPHE	12815
169	YAMPA R. D/S LAKE CATAMOUNT @ HWY 131	CDPHE	12806F
174	YMP-1 [Yampa River 200m above confl Walton Creek]	CSS[1]	YMP-1
175	YMP-2 [Yampa River 35m above confl Fish Creek]	CSS[1]	YMP-2
176	YMP-3A [Yampa River 70m above pedestrian bridge and hot spring outflow in Weiss Park]	CSS[1]	YMP-3A

Appendix 6. Description of stream sites in the Upper Yampa River watershed, Colorado, that have macroinvertebrate data, and period of water-quality record and number of sample days, 1975 through 2008.—Continued

[No , number; USGS, U.S. Geological Survey; CDPHE, Colorado Department of Public Health and Environment; CSS, City of Steamboat Springs. Subwatershed definitions: Yampa River subwatershed 1, Yampa River and tributaries upstream from Chuck Lewis State Wildlife Area; Yampa River subwatershed 2, Yampa River and tributaries from Chuck Lewis State Wildlife Area to Elk River confluence; Elk River subwatershed, Elk River and tributaries; Yampa River subwatershed 3, Yampa River and tributaries from Elk River confluence to Town of Hayden; Yampa River subwatershed 4, Yampa River and tributaries from Town of Hayden to Elkhead Creek confluence; Elkhead Creek subwatershed, Elkhead Creek and tributaries. Sites with the same site number are considered to be at the same location]

Site no. (see figure 18)	Latitude	Longitude	Subwatershed	Period of water-quality record (month/year)	No. of sample days
143	40.399	−106.834	Yampa River subwatershed 1	8/1975	1
143	40.399	−106.834	Yampa River subwatershed 1	9/1998, 4/2001, 8/2001	3
145	40.492	−106.948	Yampa River subwatershed 2	8/1975	1
147	40.269	−106.881	Yampa River subwatershed 2	4/2001, 8/2001	2
150	40.502	−107.264	Yampa River subwatershed 3	9/1975	1
151	40.479	−107.013	Yampa River subwatershed 3	4/2001, 8/2001	2
152	40.238	−106.940	Yampa River subwatershed 1	8/1975	1
153	40.483	−106.832	Yampa River subwatershed 2	8/1975	1
153	40.483	−106.832	Yampa River subwatershed 2	9/1998, 4/2001	2
154	40.488	−107.160	Yampa River subwatershed 3	9/1975	1
155	40.492	−107.296	Yampa River subwatershed 3	9/1975	1
162	40.183	−106.915	Yampa River subwatershed 1	4/2001, 8/2001	2
169	40.375	−106.825	Yampa River subwatershed 1	4/2001, 8/2001	2
174	40.449	−106.820	Yampa River subwatershed 2	9/2005, 9/2007, 9/2008	[2]3
175	40.466	−106.830	Yampa River subwatershed 2	9/2005, 9/2007, 9/2008	[2]3
176	40.481	−106.828	Yampa River subwatershed 2	9/2005, 9/2007, 9/2008	[2]3

[1]Data collected by GEI Consultants, Inc.

[2]Four samples were collected on the same day.

Publishing support provided by:
Denver Publishing Service Center, Denver, Colorado

For more information concerning this publication, contact:
Director, USGS Colorado Water Science Center
Box 25046, Mail Stop 415
Denver, CO 80225
(303) 236-4882

Or visit the Colorado Water Science Center Web site at:
http://co.water.usgs.gov/

This report is available at: http://pubs.usgs.gov/sir/2012/5214

Bauch and others—**Water-Quality Assessment and Macroinvertebrate Data for the Upper Yampa River Watershed, Colorado, 1975–2009**—SIR 2012–5214